I0485569

Geology of the Alder Creek Mining District of Idaho

by United States Geological Survey

with an introduction by Kerby Jackson

This work contains material that was originally published by
the U.S. Geological Survey in 1968.

This publication was created and published for the public benefit,
utilizing public funding and is within the Public Domain.

This edition is reprinted for educational purposes
and in accordance with all applicable Federal Laws.

Introduction Copyright 2015 by Kerby Jackson

Introduction

It has been years since the United States Geological Survey released his important publication "Geology of the Alder Creek Mining District". First released in 1968, this important volume has now been out of print for this days and has been unavailable to the mining community since those days, with the exception of expensive original collector's copies and poorly produced digital editions.

It has often been said that "*gold is where you find it*", but even beginning prospectors understand that their chances for finding something of value in the earth or in the streams of the Golden West are dramatically increased by going back to those places where gold and other minerals were once mined by our forerunners. Despite this, much of the contemporary information on local mining history that is currently available is mostly a result of mere local folklore and persistent rumors of major strikes, the details and facts of which, have long been distorted. Long gone are the old timers and with them, the days of first hand knowledge of the mines of the area and how they operated. Also long gone are most of their notes, their assay reports, their mine maps and personal scrapbooks, along with most of the surveys and reports that were performed for them by private and government geologists. Even published books such as this one are often retired to the local landfill or backyard burn pile by the descendents of those old timers and disappear at an alarming rate. Despite the fact that we live in the so-called "Information Age" where information is supposedly only the push of a button on a keyboard away, true insight into mining properties remains illusive and hard to come by, even to those of us who seek out this sort of information as if our lives depend upon it. Without this type of information readily available to the average independent miner, there is little hope that our metal mining industry will ever recover.

This important volume and others like it, are being presented in their entirety again, in the hope that the average prospector will no longer stumble through the overgrown hills and the tailing strewn creeks without being well informed enough to have a chance to succeed at his ventures.

Kerby Jackson
Josephine County, Oregon
October 2015

CONTRIBUTIONS TO ECONOMIC GEOLOGY

GEOLOGY OF PART OF THE ALDER CREEK MINING DISTRICT, CUSTER COUNTY, IDAHO

By W. H. Nelson and C. P. Ross

ABSTRACT

In the Alder Creek mining district, stocks of quartz monzonite, granite, leucogranite porphyry and related dikes, and volcanic rocks, all of early Tertiary age, seem to be differentiates of a single parent magma. The stocks have metamorphosed sedimentary rocks of the Copper Basin Formation and the White Knob Limestone, both from Early Mississippian to Early Permian in age; limestones have been in part isochemically metamorphosed to marble, and in part metasomatized to skarn. Valuable concentrations of copper, lead, and zinc minerals occur in the skarn, and mostly within skarn at the contact between the leucogranite porphyry and limestone.

Glacial deposits of two ages and alluvial deposits of two ages occur in the report area.

INTRODUCTION

Metamorphic rocks at the contact between sedimentary and intrusive rocks are magnificiently exposed in the cliffs, ridges, and cirques of the Alder Creek mining district. The sedimentary rocks are limestone, siltstone, and quartzite, and some mixed argillaceous silty and limy rocks. The intrusive rocks are a variety of compositions—all possibly differentiates of a single parent magma. Of the metamorphic rocks, skarn is of greatest interest because it contains most of the copper, lead, and zinc ores that are found in the district. Most of the concentrations of ore minerals are confined to skarn that is adjacent to the leucogranite porphyry, one of the varieties of intrusive rock.

The present summary of data on the Alder Creek mining district is a product of the observations by the authors plus those of T. H. Kiilsgaard, and published data cited below. Nelson constructed the detailed areal map (pl. 1) of most of the productive part of the district, made observations and formulated ideas as to the origin of the ore deposits during the summer of 1958. The mines were inactive during the summer of 1958, and Nelson did not visit the underground workings; control for the parts of the cross sections through the Empire mine on plate 1 was obtained from detailed mine maps made by Farwell and Full (1944). Nelson did the petrography of the igneous rocks and wrote most of the text. T. H. Kiilsgaard studied the mines briefly in 1961, and ideas he gained at that time have been incorporated here. Ross visited the district at intervals from 1929 through 1961 and saw large parts of the mine workings.

Plate 1 covers about 45 square miles of the Alder Creek mining district, west of the town of Mackay, Custer County, Idaho. The district extends locally somewhat more than a mile south of the area shown. The location of the area of this report is shown in figure 1.

SEDIMENTARY ROCKS

COPPER BASIN FORMATION

The name Copper Basin Formation was introduced by Ross (1962) for the thick assemblage of dominantly noncarbonate clastic rocks in the southern part of the Copper Basin depression. In the area shown on plate 1 the rocks of the Copper Basin Formation range from shale to conglomerate, but most of them are argillaceous siltstone and fine-grained quartzite which are medium gray to very dark gray on fresh surfaces. Much of the dark color of these rocks is due to disseminated carbon; where the rocks are weathered, the color is mostly brown because of included iron oxide.

The Copper Basin Formation and the White Knob Limestone occupy about the same stratigraphic interval; and the Copper Basin Formation, which makes up most of this interval to the west, and the White Knob Limestone, which makes up most of the interval to the east, interfinger in the vicinity of the area of this report. The alternation of rocks that are assigned to these two formations east of the Middle Fork of Navarro Creek is part of this interfingering. Ross (1960, 1962) described the regional relations of these formations, and Skipp (1961) has described interfingering between the various rocks in an area 2 miles west of that shown on plate 1.

Fossils are not abundant enough in the Copper Basin Formation to establish its age range. The formation interfingers with, and therefore is, in part at least, the same age as the White Knob Limestone, which ranges from Early Mississippian to Early Permian in age (Ross, 1962).

WHITE KNOB LIMESTONE

Ross (1962) introduced the name White Knob Limestone for the rocks that had previously been called Brazer in the vicinity of the area shown on plate 1. The name is taken from the White Knob Mountains, which include the mountains in the area of this report.

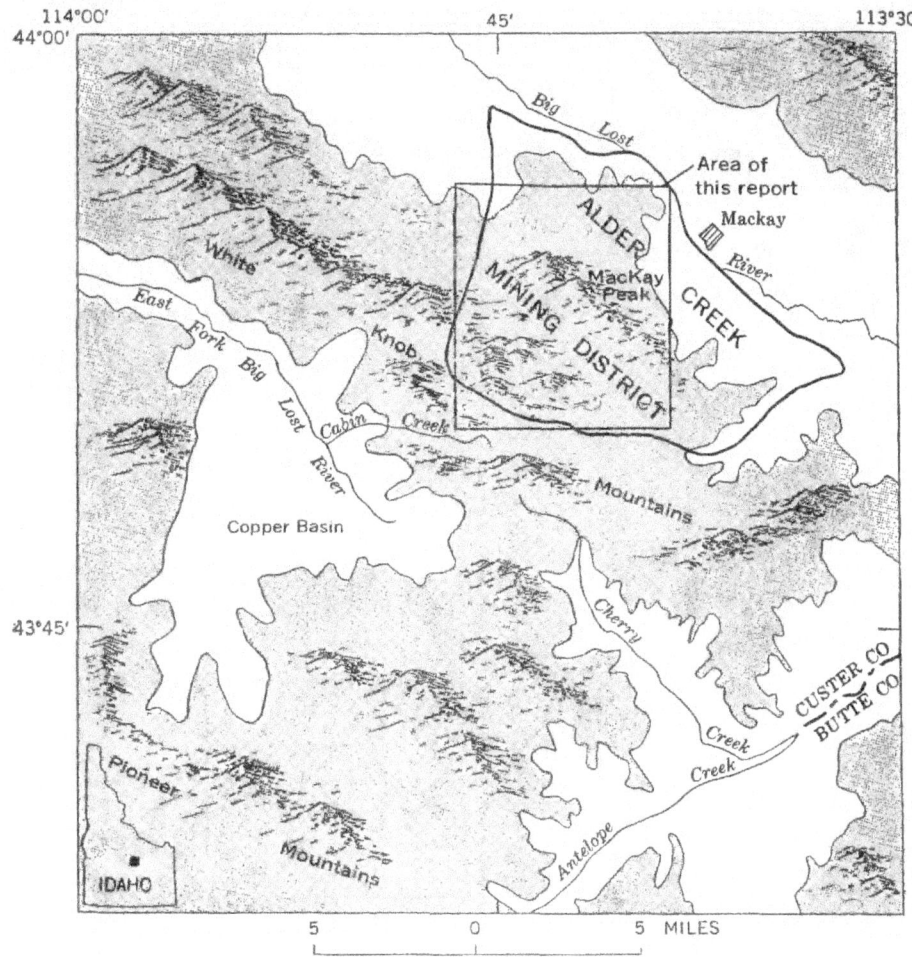

FIGURE 1.—The location of the area of this report.

Most of the White Knob Limestone is very pure limestone. It is light gray to dark gray and is in beds that range from a few inches to about 10 feet in thickness. A thin veneer of clay is common on bedding planes, and at a few places there are shaly beds several inches thick. Locally chert, in the form of nodules and beds an inch or two thick, makes up as much as 15–20 percent of the rock. Chemical analyses of two specimens typical of the limestone are given in table 1. Umpleby (1917, p. 59) presented four analyses of limestone, one of which contains 12.72 percent magnesium oxide. The magnesium content indicates that locally the limestone is dolomitic; however, Umpleby's specimens came from mine workings, and the magnesium could have been introduced locally during metamorphism instead of being a component of the original carbonate beds.

TABLE 1.—*Chemical and normative analyses of rocks of the area*

Chemical analyses

[Analyses 1, 2, and 6 by Norman Davidson; 3–5, 7–10, 12, and 13 by Margaret C. Lemon; 11 by Faye H. Neuerburg]

	1	2	3	4	5	6	7	8	9	10	11	12	13
	Quartz monzonite	Quartz monzonite	Porphyritic rhyolite	Porphyritic rhyolite	Mackay Granite	Mackay Granite	Mackay Granite	Leuco-granite porphyry	Leuco-granite porphyry	Rhyolite	Challis Volcanics	Limestone	Limestone
	Full 1	Full 2	58 N. 85	58 N. 76	58 N. 14	Full 3	58 N. 4	58 N. 87	58 N. 32	58 N. 71	CPR 648	59 N. 1	59 N. 2
SiO_2	58.09	61.85	65.99	67.14	72.49	72.57	74.42	70.16	70.41	76.49	64.43	4.77	5.47
Al_2O_3	15.14	16.06	14.73	14.68	13.86	13.31	12.97	14.73	14.56	12.52	16.85	.17	.12
Fe_2O_3	1.87	1.08	1.44	.85	.90	.99	.80	.21	.24	.70	1.99	.03	.09
FeO	3.84	4.07	1.96	2.13	.99	1.55	1.11	.64	.63	.32	1.22	.06	.03
MgO	3.90	1.99	1.42	1.38	.45	.62	.30	.72	.80	.06	1.82	.48	.39
CaO	5.74	4.70	2.73	2.88	1.35	1.74	.78	3.68	3.46	.43	3.27	52.58	52.34
Na_2O	5.00	3.54	3.87	3.31	4.25	4.90	3.73	3.31	3.29	4.08	4.20	.01	.02
K_2O	2.88	3.52	3.98	4.34	4.64	2.90	4.93	5.45	5.41	4.58	3.05	.03	.02
H_2O+	1.37	1.25	1.67	1.24	.35	.53	.28	.22	.26	.38	1.38	.05	.11
H_2O-	.24	.24	.41	.31	.09	.09	.12	.09	.23	.11	.40	.04	.02
TiO_2	.90	.83	.53	.45	.24	.26	.19	.34	.36	.04	.45	.01	.01
P_2O_5	.38	.13	.19	.20	.07	.11	.05	.14	.14	.03	.22	.01	.03
MnO	.15	.20	.03	.03	.04	.09	.08	.04	.01	.16	.07	.02	.01
CO_2	.04	.74	1.13	.75	.01	.10	.02	.02	.04	.01	1.28	41.53	41.31
Cl			.01	.03	.06		.09	.09	.04			.00	.00
F		.05	.07	.06	.14	.08	.07	.05	.04	.09		.01	.01
Subtotal	------	100.25	100.20	99.78	99.93	99.84	99.94	99.89	99.92	100.04	------	99.83	99.98
Less O	------	.02	.03	.04	.07	.03	.05	.04	.03	.04	------	.00	.00
Total	99.54	100.23	100.17	99.74	99.86	99.81	99.89	99.85	99.89	100.00	99.63	99.83	99.98

C.I.P.W. norm

[Calculated by Burroughs B220 computer]

q	2.445	14.759	23.413	24.848	27.092	28.566	32.075	23.499	23.984	35.085	24.236	
or	17.015	20.796	23.514	25.641	27.413	17.133	29.127	32.199	31.963	27.059	18.020	
ab	42.286	29.938	32.655	27.771	35.499	41.440	30.880	27.328	27.528	34.431	35.520	
an	10.372	17.461	4.707	7.864	5.173	5.768	2.920	9.599	9.147	.277	6.692	
c		.030	2.348	1.701	.040		.560			.767	4.190	
hl			.016	.049	.099		.148	.148	.066	.016		
wo	6.415					.401		3.045	2.835			
en	9.709	4.954	3.535	3.436	1.120	1.544	.747	1.792	1.992	.149	2.041	
fs	4.298	5.582	1.664	2.521	.752	1.766	1.212	.515	.438	.018		
mt	2.711	1.566	2.088	1.232	1.305	1.435	1.160	.304	.348	1.015	2.856	
hm											.020	
il	1.709	1.576	1.007	.855	.456	.494	.361	.646	.684	.076	.855	
ap	.900	.308	.450	.474	.166	.261	.118	.332	.332	.071	.521	
fr		.091	.126	.105	.281	.154	.139	.090	.009	.182		
cc	.091	1.683	2.570	1.706	.023	.227	.045	.045	.023	.364	2.911	
Total	97.951	98.744	98.093	98.203	99.419	99.189	99.492	99.542	99.409	99.510	97.862	

Molecular norm

Q	2.25	13.59	22.00	23.23	24.86	26.67	29.61	21.46	22.24	32.37	23.12	
Or	17.15	21.15	24.00	26.25	27.70	17.30	29.55	32.50	32.25	27.35	18.40	
Ab	45.30	32.30	35.40	30.40	38.55	44.45	34.00	30.00	29.85	37.05	38.40	
An	10.45	17.80	5.25	8.45	5.10	5.85	3.35	8.40	9.15	.95	6.25	
C			2.48	1.69			.32			.57	4.89	
Wo	6.22	.52			.44	.58		3.36	2.90			
En	10.86	5.58	4.00	3.90	1.26	1.74	.84	2.00	2.22	.16	2.30	
Fs	3.64	4.76	1.22	2.18	.64	1.48	1.02	.41	.38	.52		
Mt	1.97	1.16	1.63	.90	.95	1.04	.84	.23	.26	.75	2.10	
Hm											.02	
Il	1.26	1.18	.75	.62	.34	.38	.28	.48	.50	.04	.64	
Ap	.80	.28	.40	.43	.16	.24	.14	.16	.29	.06	.48	
Co	.10	1.48	2.92	1.94	.02	.26	.06	.06	.02	.40	3.52	
Total	100.00	99.80	100.05	99.99	100.02	99.99	100.01	99.06	100.06	100.22	100.12	

See end of table for description of localities.

TABLE 1.—*Chemical and normative analyses of rocks of the area*—Continued

Semiquantitative spectrographic analyses

[Determined spectrographically by Paul R. Barnett. Looked for but not detected: Ag, As, Au, B, Bi, Cd, Ge, In, Pt, Sb, Ta, Th, Tl, U, W, and Zn. Data have an overall accuracy of ±15 percent except that they are less accurate near limits of detection, where only one digit is reported]

	1	2	3	4	5	6	7	8	9	10	11	12	13
	Quartz monzonite	Quartz monzonite	Porphyritic rhyolite	Porphyritic rhyolite	Mackay Granite	Mackay Granite	Mackay Granite	Leucogranite porphyry	Leucogranite porphyry	Rhyolite	Challis Volcanics	Limestone	Limestone
	Full 1	Full 2	58 N. 85	58 N. 76	58 N. 14	Full 3	58 N. 4	58 N. 87	58 N. 32	58 N. 71	CPR 648	59 N. 1	59 N. 2
Cu			0.0008	0.0001	0.0003		0.0006	0.004	0.018	0.0001		<0.0002	<0.0002
Pb			.004	.004	.004		.009	.005	.005	.004		<.002	<.002
Mo			<.0002	<.0002	<.0003		<.0003	.0006	.0012	<.0002		<.0004	<.004
Ba			.12	.15	.073		.042	.14	.070	.002		.004	.003
Be			.0005	.0004	.0007		.0006	.0004	.0005	.0020		<.0004	<.0004
Co			.0005	.0004	.0003		.0003	<.0003	<.0003	<.0002			
Cr			.0022	.0036	.0006		.0005	.0012	.0014	.0001		.0010	.0010
Ga			.0016	.0014	.0016		.0014	.0014	.0012	.0017		.001	.001
La			.007	.006	.008		.008	.008	.008	<.005		<.01	<.01
Nb			.003	.002	.006		.008	.04	.04	.007		<.003	<.003
Ni			.0010	.0010	<.0003		<.0003	.0005	.0006	<.0002		<.004	<.004
Sc			.0008	.0008	.0008		.0008	.0008	.0008	.0005		<.001	<.001
Sn			<.0003	.0007	.0005		.0008	.0005	.0014	.0008		<.0006	<.0006
Sr			.030	.050	.022		.014	.036	.038	.005		.07	.05
V			.0046	.0041	.0018		.0011	.0034	.0031	<.0005		<.001	<.001
Y			.002	.002	.004		.003	.002	.002	.004		<.003	<.003
Yb			.0002	.0002	.0004		.0004	.0001	.0002	.0005		<.0002	<.0002
Zr			.019	.020	.032		.028	.026	.032	.0094		.004	.004

1. Ridge north of Horseshoe mine.
2. Empire mine, 1,600-ft level.
3. 3,000 ft N. 18° W. of Grand Prize mine.
4. 2,800 ft N. 38° W. of Horseshoe mine.
5. 2,900 ft S. 17° W. of the top of White Knob.
6. Mackay Peak.
7. 1,800 ft N. 45° E. of the top of Mackay Peak.
8. 2,800 ft N. 18° W. of Grand Prize mine.
9. 2,800 ft N. 10° W. of Grand Prize mine.
10. 5,650 ft S. 6° E. of the top of White Knob.
11. 88,500 ft S. 78° W. of the top of Mackay Peak.
12. 2,750 ft N. 49° E. of Grand Prize mine.
13. 750 ft S. 56° E. of Blue Bird mine.

The noncarbonate sedimentary rocks within the White Knob range from shale to conglomerate, but most of them are argillaceous siltstone and fine-grained quartzite. These rocks are medium gray to very dark gray on fresh surfaces. Much of the dark color of these rocks is due to disseminated carbon; where weathered, much of the rock is brown because of included iron oxide.

Small masses of siltstone, quartzite, shale, and conglomerate similar to the rocks that make up the Copper Basin Formation occur in the White Knob Limestone. As noted, the Copper Basin Formation and White Knob Limestone interfinger in this area, and these bodies of siltstone, quartzite, shale, and conglomerate are believed to be either detached pods and lenses of rocks, such as those that make up the Copper Basin Formation, or parts of thin wedges of the Copper Basin Formation that extend eastward into the White Knob Limestone.

Fossil evidence indicate that the White Knob Limestone ranges from Early Mississippian to Early Permian in age (Ross, 1962, p. 385).

INTRUSIVE ROCKS

All of the intrusive rocks and perhaps some of the extrusive rocks may have been derived by differentiation from a single parent magma. There is no direct evidence of the existence or composition of such a parent magma; however, the variation diagram (fig. 2), coupled with what can be deduced from field relationships about the relative ages of the various intrusive rocks, suggests that the magma tended to become more acidic with time. It will be noted on the variation diagram (fig. 2) that the order of the leucogranite porphyry and the Mackay Granite based on their silica content is reversed from their deduced order of emplacement based on field evidence. This should not be surprising because suites of igneous rocks that are believed to have been derived from a single differentiating magma commonly depart from theoretical differentiation curves for a number of reasons.

Table 1 gives the compositions of the igneous rocks. The intrusive rocks, except for the dikes, are described below in their probable order of emplacement. The dikes, some of which seem to have been intruded over a considerable span of time, are described after the other intrusive rocks.

The granitic rocks are clearly intrusive. Their relationship to the enclosing sedimentary rocks is especially evident in the vicinity of White Knob where the contact locally follows bedding; also at numerous places, dikes of granite rock extend out into cracks in the sedimentary rocks.

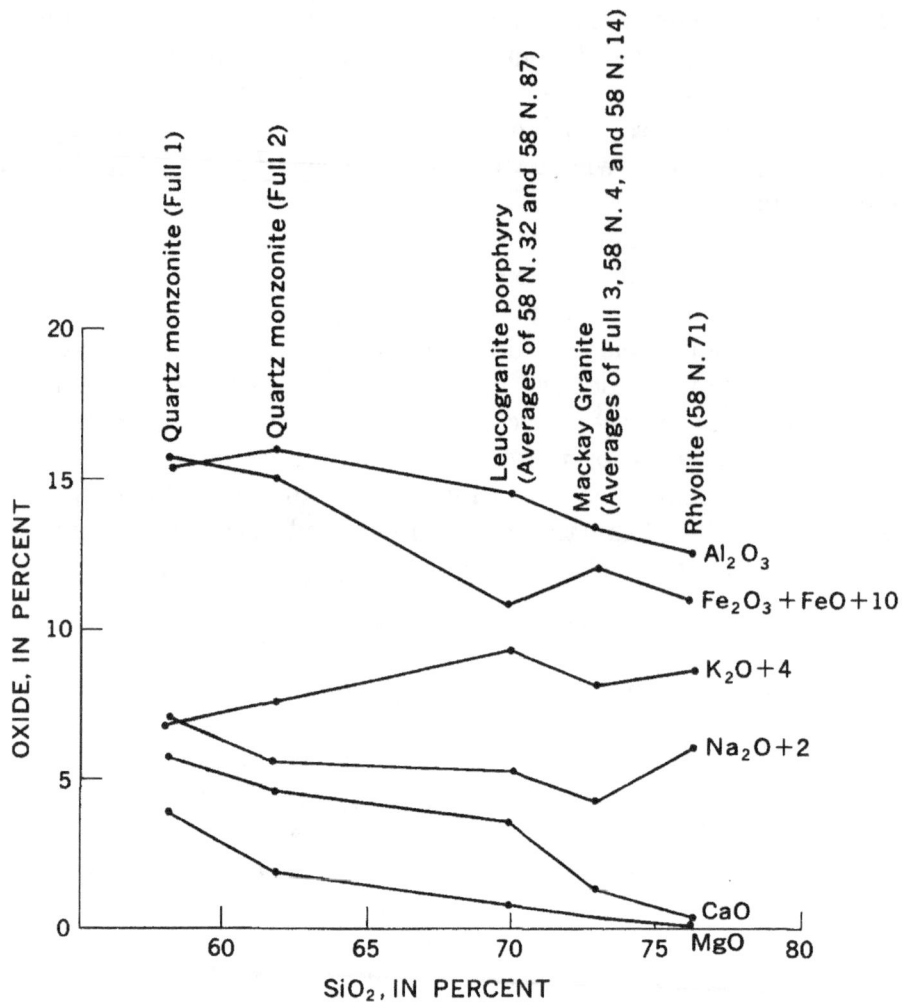

FIGURE 2.—Variation of selected oxides compared to variation in silica. Data from table 1.

QUARTZ MONZONITE

The rock termed quartz monzonite in this report is exposed in three small areas north of the Mackay Granite and in some of the workings of the Empire mine. It has a wider range of composition and appearance than the other intrusive rocks of the area, with the exception of the dikes. That the most common rock types, as well as the range of rock types, seems to be about the same in all these bodies suggests a close relationship between the rocks in the various quartz monzonite bodies. An additional reason for believing that each and perhaps all of these bodies of quartz monzonite are the result of a single episode of intrusion is the gradation of various textures and compositions within these bodies.

The description that follows applies specifically to the stock that is in contact with the Mackay Granite on the northeast side of Mackay Peak. But as noted, above, the rocks of all the quartz monzonite stocks seem to be comparable.

The quartz monzonite ranges from light gray to medium gray and is, on the average, somewhat darker and finer grained than the Mackay Granite to be described below. Most of the groundmass, which makes up from 40 to 50 percent of the quartz monzonite, is fine grained, but at a few places it is medium grained; it consists of a mixture of orthoclase, quartz, and a little oligoclase; the orthoclase is slightly more abundant than the quartz. In most of the stock the quartz is confined to the groundmass, but locally as much as one-fourth of it is in phenocrysts, which are slightly larger than the grains of the groundmass, but considerably smaller than the feldspar pheno- crysts. These small quartz phenocrysts are inconspicuous in hand specimens; they are rounded and embayed. Orthoclase is confined to the groundmass of most of these rocks, but in a few places it occurs as phenocrysts. Oligoclase phenocrysts make up about 25 percent of these rocks and are as much as 1.5 mm in length. Mafic minerals con- tribute 10–20 percent to the volume of the rock and include diopside, green hornblende, and biotite either singly or in combination. The hornblende is commonly associated with and in part rims the diopside. The biotite commonly includes magnetite, and, in some specimens is partly altered to chlorite. Accessory minerals are apatite, sphene, and magnetite.

Farwell and Full (1944, p. 7) called these rocks diorite, and they undoubtedly range from diorite through quartz monzonite; how- ever, most of them cannot be classified as diorite because the ratio of orthoclase to oligoclase is too great. We believe most of the rocks contain enough quartz to be classified as quartz monzonite rather than monzonite. Table 1 includes a chemical and a normative analysis of one of the less silicic of these rocks.

A horizontal thickness of at least 370 feet of quartz monzonite oc- curs in the Empire mine on the 1,600-foot level (Farwell and Full, 1944, pl. 23). This level was not accessible during the present study, but a few specimens collected by Farwell and Full were available for study. Although these specimens are coarser grained and contain more quartz than most of the quartz monzonite exposed at the sur- face, they resemble the surface rocks sufficiently in mineralogy and position relative to the Mackay Granite to suggest that they are related to the surface rocks. The amount of quartz in these specimens is within the range observed in the quartz monzonite bodies at the surface. The grains in these specimens range from 0.2 to 10 mm; the most common size is a little more than 1 mm. Locally some of the feldspar crystals are enough larger than the rest of the mineral grains to be termed phenocrysts. These specimens are composed of 10–15 percent quartz, about 45 percent orthoclase, 30 percent oligoclase, and about 10 percent mafic minerals, mostly green hornblende. Chemical and normative analyses of a specimen of the quartz monzonite from the 1,600-foot level of the Empire mine are included in table 1.

There is little evidence to indicate the age of the quartz monzonite relative to the other intrusive rocks of the area. Along the ridge south of the Horseshoe mine, rocks that are devoid of megascopically identifiable quartz are in abrupt contact with rocks that contain numerous quartz phenocrysts. The former rocks seem to be part of a quartz monzonite body, and the latter seem to be dikes that extend out from the Mackay Granite. These relationships suggest that the quartz monzonite is older than the Mackay Granite.

MACKAY GRANITE

The largest of the intrusive bodies in the area of this report (pl. 1) is here named the Mackay Granite after excellent exposures at its type locality, Mackay Peak (SW¼ sec. 35, T. 7 N., R. 23 E.). This name is a local term which has long been informally applied to this intrusive mass. Although the local name is adopted in this report, it should be noted that the rock is actually granite porphyry. It is exposed over an area of about 11½-square miles. This granite is fairly uniform in texture throughout most of its exposed extent, but shows local variations which seem to be related to the distance from the outer edge of the mass at the time of emplacement.

Most of the Mackay Granite is somewhat weathered and is pinkish gray; less weathered granite, which has been exposed by glacial scouring, is commonly a faintly greenish gray. The bulk of the rock is a porphyry made up of phenocrysts of orthoclase and quartz in a fine-grained groundmass. The orthoclase phenocrysts make up 25–35 percent of the granite; they are as much as 10 mm in diameter, subhedral to euhedral, and pinkish gray in hand specimen. In thin section many of them are seen to enclose small rounded anhedral quartz inclusions. The quartz phenocrysts make up 5–15 percent of the rock, are medium dark gray in hand specimen, as large as 5 mm in size, rounded, and are embayed by and contain rounded inclusions of groundmass material. The groundmass is a mosaic of nearly equidimensional anhedral grains of quartz, orthoclase, and plagioclase. The orthoclase and the quartz of the groundmass each make up 20–30 percent of the rock, and the grains range from about 0.05 to 0.3 mm in size. The plagioclase is albite-oligoclase in composition and makes up 10–15 percent of the granite. Biotite and hornblende each occur in amounts of 1 or 2 percent and are locally altered to chlorite; magnetite, apatite, zircon, and rutile are present in amounts of less than 1 percent. Three chemical analyses of typical Mackay Granite are given in table 1.

A specimen of the Mackay Granite collected by Farwell and Full from the 1,600-foot level (about 7,010 ft above sea level) of the Empire mine has phenocrysts similar to the typical granite and has the same composition, but the grains of the groundmass are larger and range from 0.15 to 1.5 mm in size.

Locally, especially along the ridges northeast of the head of Corral Creek and north of the head of Steward Canyon, some of the Mackay Granite has a finer grained groundmass and a deeper greenish color than the rest of the Mackay Granite. This rock is believed to be a chilled phase of the granite, which is more deeply colored because more hornblende and biotite have been altered to chlorite than in the typical granite.

T. W. Stern (written commun., Sept. 21, 1959) reported that two specimens of Mackay Granite have ages of 40±10 million years as determined by the lead-alpha method, using zircons. Thus the age of the Mackay Granite is considered to be early Tertiary.

LEUCOGRANITE PORPHYRY

The leucogranite porphyry is confined to an irregular area along the northeast side of the Mackay Granite and is in contact with the granite. The rocks of this mass are very light gray and porphyritic. The groundmass is composed of grains so small that the proportions of minerals in it are very difficult to estimate. Chemical analyses of this rock (table 1) show that it can appropriately be termed granite, or more specifically, because of its very light color and porphyritic texture, leucogranite porphyry.

From one-third to two-thirds of the leucogranite porphyry is groundmass made up of fine-grained quartz and feldspar; most of the feldspar is orthoclase. This groundmass is commonly a mosaic of almost equidimensional grains, but locally it has a randomly oriented fibrous texture. The groundmass grains range from 0.007 to 0.06 mm in size, but the full range of size is not present in any one specimen. Orthoclase and quartz phenocrysts occur in about equal amounts, and each is usually somewhat more abundant than oligoclase phenocrysts. Orthoclase phenocrysts are as large as 20 mm, oligoclase phenocrysts as large as 4 mm, and quartz phenocrysts as large as 8 mm in size. The phenocrysts are euhedral to subhedral, and the quartz phenocrysts commonly have rounded corners and are embayed by and contain rounded inclusions of groundmass material. Diopside is the only significant mafic mineral in these rocks, it contributes only 3–5 percent to the volume of the rock and occurs as widely scattered small anhedral grains and in aggregates, which also contain orthoclase. The diopside in these aggregates occurs either as randomly oriented, nearly equidimensional anhedral grains or as rectangular or rodlike grains, which in a single aggregate have a common orientation and seem to be skeletal crystals. The orthoclase in these aggregates occurs as nearly equidimensional, randomly oriented anhedral grains. In some specimens calcite and sphene also make up a small part of the volume of these diopside-bearing aggregates. Table 1 includes chemical and normative analyses of two typical specimens of the leucogranite porphyry.

Umpleby (1917, p. 60) reported various stages of replacement of hornblende and biotite by diopside in granitic rocks associated with skarn. We did not observe similar features in the leucogranite porphyry, and it is difficult to evaluate the significance of these observations by Umpleby because he was not explicit about their extent or location.

Evidence of the age of the leucogranite porphyry relative to the Mackay Granite is equivocal. Exposed contacts between these rocks are scarce, and most are in mine workings. Where the contracts are exposed, rocks resembling the Mackay Granite have various relationships with rocks resembling the leucogranite porphyry. Interpretation of these relationships is complicated by difficulty in positively identifying the rocks of the various plutons and dikes in small exposures.

We believe that the leucogranite porphyry is younger than the Mackay Granite because it resembles the rhyolite dikes which cut the Mackay Granite. At several places, notably in the area a short distance south of the White Knob mine, the leucogranite porphyry grades at its margins into rhyolite which seems to be identical with rhyolite in the dikes that cut the Mackay Granite. A diligent search did not reveal any rhyolite dikes within the leucogranite porphyry, and this, along with the fact that the leucogranite porphyry is athwart the dike zone, tends to confirm the interpretation that the rhyolite dikes are related to and were emplaced at the same time as the leucogranite porphyry.

Minor variations of composition and texture within the leucogranite porphyry masses, especially variations in the abundance of quartz phenocrysts, suggest that the leucogranite porphyry may have been emplaced during several episodes of intrusion, but even so it seems likely that all the leucogranite porphyry was emplaced during a fairly brief time interval.

Relations which suggest the opposite order of emplacement were observed by T. H. Kiilsgaard (written commun., Jan. 19, 1962) on the ridge about 2,200 feet northeast of Mackay Peak. At that place a small dike that seems to be an apophysis of the Mackay Granite intrudes a leucocratic granitic rock. Because the quantity of quartz phenocrysts in this leucocratic rock is at the lower limit of the range of quartz phenocrysts in the leucogranite porphyry, this leucocratic rock may not be directly related to the greater part of the leucogranite porphyry.

T. W. Stern (written commun., Sept. 21, 1959) determined a specimen of the leucogranite to be 50 ± 10 million years old by the lead-alpha method, using zircons. These figures overlap the age range determined by Stern for the Mackay Granite and therefore do not conclusively establish the relative ages of the two rocks.

Umpleby (1917, p. 44, 94) considered the rocks we call leucogranite porphyry to be merely a border phase of the Mackay Granite, which he called granite porphyry. We believe, however, that the differences between the leucogranite porphyry and the Mackay Granite are too consistent and the contact between them too abrupt for Umpleby's interpretation to be valid.

Kemp and Gunther (1908, p. 275) believed that their quartz porphyry, which seems to be equivalent to our leucogranite porphyry, was formed later than the main mass of granite, the Mackay Granite of the present report.

QUARTZ LATITE

Quartz latite ranges from light greenish gray to dark greenish gray and occurs as dikes. Locally some of it weathers to a light to very light yellowish gray similar to the color of most of the rhyolite described below, but it can be distinguished from the rhyolite by the absence of quartz grains large enough to be seen in hand specimens. The dikes of quartz latite are much more heterogeneous in color and texture than the rhyolite dikes; however, all the quartz latite dikes and the similarly heterogeneous quartz monzonite bodies are believed to be related because they all lack quartz phenocrysts and many of the quartz latite dikes originate from and grade into the quartz monzonite stocks. These rocks are called quartz latite on the basis of their assumed relationship with the quartz monzonite.

A very fine grained groundmass makes up more than three-quarters of the volume of all the quartz latite. The groundmass consists of orthoclase, oligoclase, and quartz. Orthoclase is more abundant than oligoclase, and quartz is much less abundant than either. Most of the feldspar grains are elongated prisms, which are commonly arranged in radiating clusters. The quartz occurs as small isolated grains interspersed among the feldspar grains. Small phenocrysts of altered oligoclase make up a small part of the volume of all these rocks, and small patches of chlorite mixed with calcite, magnetite or limonite, and commonly a little quartz, attest to the former presence of small mafic phenocrysts. Locally a little pyroxene, amphibole, or biotite remains.

The quartz latite dikes that originate from and grade into the quartz monzonite stocks are the same age as the quartz monzonite, which is probably the oldest intrusive rock in the area. Locally, quartz latite dikes are crossed by rhyolite dikes (northeastern part of pl. 1), and this tends to confirm the interpretation that the quartz latite is among the oldest igneous rocks in the area.

RHYOLITE

Rhyolite ranges from very light gray to medium gray; most of it has a faint yellowish tint, but some of it is greenish. All of it contains euhedral to subhedral dipyramid phenocrysts of high temperature quartz, and in many specimens some of the quartz is smoky. Oligoclase and orthoclase together or individually are common as phenocrysts. Most of the phenocrysts do not exceed 5 mm in size. Brownish patches 2–5 mm across occur in most of the rhyolite; these patches consist of mixtures of chlorite, limonite, cryptocrystalline quartz(?), and locally biotite. Biotite occurs as isolated fresh grains in a few specimens, but more commonly it is partly or completely altered to chlorite. Calcite occurs as an alteration product in oligoclase and is locally scattered through the groundmass. A very fine grained groundmass makes up more than 75 percent of all the rhyolite, and appears to consist dominantly of a mixture of quartz and feldspar. Table 1 gives the chemical composition of a typical rhyolite. As previously noted, we believe that the rhyolite, which occurs as dikes, is related to the leucogranite porphyry, and therefore is the same age. Section A–A', plate 1, shows our interpretation of the relationship of the rhyolite dikes to the leucogranite porphyry.

In the area north of the Blue Bird mine, many of the rhyolite dikes are locally thicker than elsewhere in the area, and these thicker parts of the dikes contain large feldspar phenocrysts similar to those in dikes that are called porphyritic rhyolite in this report. In general the groundmass of the thicker rhyolite dikes that contain large feldspar phenocrysts is lighter colored than the groundmass of the porphyritic rhyolite. At least locally where porphyritic rhyolite dikes cut rhyolite dikes the porphyritic rhyolite is younger than the rhyolite.

PORPHYRITIC RHYOLITE

Porphyritic rhyolite, which occurs only as dikes, differs from most of the other dike rocks of the area in that it contains abundant large feldspar phenocrysts. Three-quarters or more of the volume of all the porphyritic rhyolite is fine-grained gray or greenish-gray groundmass consisting dominantly of orthoclase with shreds of chlorite and scattered quartz grains. Enclosed within the groundmass are phenocrysts of feldspar, quartz, completely altered mafic minerals, and locally, rare flakes of biotite. White euhedral phenocrysts of oligoclase occur in all the porphyritic rhyolite, and orthoclase phenocrysts occur in most of it in somewhat lesser abundance. Commonly the orthoclase is rimmed by oligoclase to form rapakivi-type phenocrysts; the rim of oligoclase may be from $\frac{1}{10}$ to $\frac{8}{10}$ of the radius of the phenocrysts. The phenocrysts of orthoclase are commonly 15 mm long, and a few are 25 mm long; very few oligoclase phenocrysts exceed about 10 mm in length. Quartz phenocrysts are rounded and embayed and have a maximum diameter of about 10 mm, but are commonly not conspicuous in hand specimens. The former presence of mafic minerals is attested to by prismatic patches of chlorite intergrown with epidote, calcite, or partly altered biotite. Table 1 gives chemical and normative analyses of two samples of the porphyritic rhyolite. Farwell and Full (1944, p. 8) used the term porphyritic granodiorite to refer to these rocks.

Dark fine-grained dikes which are similar to the groundmass of the porphyritic rhyolite occur near and most commonly at the borders of a few of the porphyritic rhyolite dikes. These dark fine-grained rocks may be a chilled phase of the porphyritic rhyolite, but the large phenocrysts in the porphyritic rhyolite may be of intertelluric origin and, if so, would be present even in a chilled phase. The fact that the abundance and size of phenocrysts varies widely in the porphyritic rhyolite suggests that the magmas for the dark dikes as well as for the porphyritic rhyolite, were derived from the magma reservoir during different stage of partial crystallization. Another possibility is that the dark fine-grained dikes are the result of the porphyritic rhyolite magma having been forced through cracks that were so narrow that the larger phenocrysts could not pass through.

The differences between the porphyritic rhyolite and the other intrusive rocks of the area are almost surely due to differences in cooling history; the porphyritic rhyolite probably remained at a temperature that was optimum for the formation of feldspars longer than the other intrusive rocks of the area.

The rounded and embayed quartz phenocrysts in the porphyritic rhyolite are identical in form to those in the Mackay Granite and the leucogranite porphyry, and therefore these three types of intrusive rocks are probably rather closely related differentiation products of a single parent magma.

The amount of variation in the porphyritic rhyolite dikes suggests that they may have been intruded over a significantly long time interval. Some of these dikes seem to be identical with chilled borders of the Mackay Granite, and therefore these dikes may be about the same age as the Mackay Granite. Some, which cut the Mackay Granite, must be younger than that part of the Mackay Granite that they cut. A few of the porphyritic rhyolite dikes cut and hence must be younger than the leucogranite porphyry. In sec. 27, T. 7 N., R. 23 E., a dike containing large feldspar phenocrysts seems to be cut by a dike without quartz or large feldspar phenocrysts, which shows that some of the porphyritic rhyolite dikes are older than some of the quartz monzonite dikes.

EXTRUSIVE ROCKS

CHALLIS VOLCANICS

In the area shown on plate 1, volcanic rocks—the Challis Volcanics—are everywhere in contact with sedimentary rocks of late Paleozoic age. In a small part of the northeast corner of the area shown on plate 1, the volcanic rocks lie on an irregular surface on limestone. In the northwestern part of the area the contact between the volcanic and sedimentary rocks lies along the base of an escarpment that is probably in part the result of faulting. The uppermost volcanic rocks here are believed to be younger than most of this faulting because they extend a short distance southward across this fault. The contact in the eastern part of the area is a fault that is locally the locus for jasperoidization.

The Challis Volcanics in the map area seems to correspond in composition to the latite-andesite member of this formation in adjacent areas (Ross, 1937, p. 51–53, and 1947, p. 1120–1121). The rocks consist of massive layers which probably include flows, ordinary indurated tuffs, and welded tuffs, as well as tuff breccias. Most of the layers are rather somber shades of brown, reddish brown, greenish gray, and gray; browns and reddish browns are most common. Locally a few of the layers are light tan. These rocks probably range from andesite to rhyolite, and rocks of latitic composition seem to be most common. Phenocrysts as much as 1 mm in length make up from about 5–35 percent of the rocks. Oligoclase makes up more than half of the phenocrysts. Augite and biotite are the most common mafic minerals in the phenocrysts, but a little hornblende is present in some of the flows. Quartz is present in some of the flows, and in a few it is as abundant as the oligoclase. Locally some of the quartz is rounded and resorbed. The groundmass is commonly an aphanitic mass that is probably dominantly feldspar. Table 1 includes a chemical analysis of one specimen of Challis Volcanics.

Rounded rock fragments are included in the Challis Volcanics just west of the road on the ridge west of Tuscarora Gulch 1,400 feet north of the south edge of the area shown on plate 1. Farwell and Full (1944, p. 9) interpreted these to be granite and concluded that the enclosing volcanic rocks were extruded after the granite was emplaced.

These rounded inclusions are composed of phenocrysts of plagioclase and biotite and locally a little quartz and sanidine in a glassy groundmass. Most of the quartz and feldspar phenocrysts are rather intensely broken, and some of the quartz is rounded and resorbed. The groundmass of these rocks is glassy, and some of it includes laminated glass fragments which look like pumice. Some of these inclusions may be xenoliths of plutonic rocks which have had their groundmasses melted and locally inflated, or as is more likely, they may all be fragments of volcanic rock derived from nearby flows. If any are xenoliths of plutonic rock, then their presence shows that the plutonic bodies from which they were derived had crystallized below the surface before the enclosing volcanic rocks were extruded.

A more definite indication of the relative age of the volcanic rocks is provided by the fact that a rhyolite dike cuts the Challis Volcanics in the northeast corner of the area shown on plate 1. The rhyolite dikes are among the youngest of the intrusive rocks of the area; so, this indication that they are younger then the volcanic rocks does not necessarily conflict with the possible presence of xenoliths of intrusive rocks in the volcanic rocks elsewhere in the area. In any case the extrusive and intrusive rocks in the area are about the same age and composition and, hence, may be genetically related.

Ross (1961, p. C179) concluded that the Challis Volcanics may range in age from Eocene to, at most, early Miocene and that probably most of the formation is Oligocene.

METAMORPHIC ROCKS

Metamorphic rocks occur near the contacts of all the larger intrusive masses in the area and along the contact between limestone and Challis Volcanics. Adjacent to the quartz monzonite and the granite the metamorphism has been dominantly isochemical; that is, recrystallization and rearrangement of chemical constituents of the rocks have taken place, under the influence of elevated temperature, without any appreciable transport of material. Adjacent to the leucogranite porphyry, on the other hand, there has been rather large-scale metasomatism, that is, introduction and removal of some of the components of the rocks, to form skarn. At the contact between the Challis Volcanics and limestone some of the rock has been metasomatized to form jasperoid.

ISOCHEMICAL METAMORPHOSED ROCKS

Isochemical metamorphosed rocks are not shown on plate 1; they form a zone several hundred feet thick at most places near the contacts of the larger intrusive masses. The zone attains its maximum width on the ridge northeast of White Knob, where marble extends about 1,000 feet to the east of the limestone-granite contact. At this place, the contact seems to follow bedding, which is nearly vertical at the surface, but the contact may dip under the limestone so that the surface exposure of marbleized limestone may be wider than the true thickness.

As previously noted the limestone is composed of relatively pure calcite, and the effect of isochemical contact metamorphism simply has been to recrystallize the calcite into larger grains. Locally the limestone contains thin shaley interbeds and beds and nodules of chert. The silica and clay minerals in these beds and nodules have reacted with the enclosing calcite to form lime silicate minerals, which occupy the same positions in the rock and have the same shape as the beds and nodules from which they formed. The lime silicate minerals are, separately or in combination, wollastonite, scapolite, diopside, tremolite, and grossularite.

At the western edge of the area shown on plate 1 the Mackay Granite intruded sandstone and siltstone rather than limestone. In these clastic rocks the effects of the contact metamorphism are much less obvious than in the limestone. The quartz grains have grown together to form mosaics, and the argillaceous components have reacted with one another and quartz to form muscovite and biotite. Near where the Mackay Granite cuts the contact between limestone and clastic rocks, on the ridge east of Corral Creek, some of the metamorphosed siltstone contains considerable diopside, which suggests that the rocks there originally contained some carbonate minerals.

SKARN

Skarn is restricted to and occurs locally at or near the contacts between limestone and all the larger intrusive masses shown on plate 1. The skarn is commonly only a few feet or less in thickness at the margins of the Mackay Granite and the quartz monzonite. All the thicker masses of skarn seem to be associated with the contact between limestone and leucogranite porphyry, especially with the margins of limestone that are partly or completely surrounded by leucogranite porphyry. A mass of skarn several tens of feet thick, between quartz monzonite and limestone on the 1,600-foot level of the Empire mine, is shown on section $B-B'$, plate 1. This is one of several pods of skarn on this level of the mine that have about the same relationship to quartz monzonite. One of the pods, shown on section C–C', is in part between quartz monzonite and limestone and in part between leucogranite porphyry and limestone; and we believe all these pods may be more closely related to nearby leucogranite porphyry than to the quartz monzonite. Some patches of skarn within the leucogranite porphyry are not now associated with limestone, and these, we believe, are all that remains of engulfed blocks, septum, and roof pendants of metasomatized limestone. The cross sections, plate 1, illustrate these relationships.

These patches of skarn within the leucogranite porphyry seem to be the basis for the statement by Kemp and Gunther (1908, p. 269, 270) "that while the deposits are associated in a general way with the contact of an eruptive rock with limestone * * * the garnetization has taken place not in the limestone, as is usually the case, but in the igneous rock itself." Umpleby (1917, p. 56, 57) also noted that the "most striking and noteworthy feature of the distribution of the garnet rock is that all the larger areas lie within the main granite porphyry mass."

The widely scattered relict bedding and the fossils within the skarn clearly show that much of the skarn has originated by metasomatism of limestone. Umpleby (1917, p. 17) noted that, "Locally the lime silicate rock has inherited in part the pattern of the igneous rock and elsewhere it shows the bedding structure of the limestone." Ross (1930, p. 15) and Farwell and Full (1944, p. 9) concurred with this view that skarn formed mostly from limestone, but also in part from the intrusive rock.

Umpleby (1917, p. 71) concluded that the skarn was formed with little or no change of volume, and we have found no evidence of volume changes that can be attributed to formation of the skarn.

Locally veins of skarn extend into the granitic rocks as much as 100 feet; the dikelike body of skarn shown on plate 1 in the cirque at the head of Mammoth Canyon is probably such a body.

The contacts between the skarn and the limestone or intrusive rocks are generally quite sharp. These contacts are best exposed in several areas where the skarn is too thin to be shown on plate 1. One such place is on the ridge 3,200 feet northeast of the junction of Mammoth and Stewart Creeks, where domical protuberances of the upper surface of granite just reach or just fail to reach the ridge crest. The relationships bteween skarn, limestone, and granite are also well exposed around the small cupola of granite 3,200 feet west-northwest of the Horseshoe mine.

The skarn consists dominantly of garnet mixed with significant quantities of pyroxene, along with subordinate amounts of magnetite, hematite, actinolite, scapolite, wollastonite, epidote, and fluorite. The garnet makes up more than three-quarters of the volume of most of the skarn. Most of the garnet is brown; chemical analyses presented by Umpleby (1917, p. 63) show that it is composed of somewhat more than half andradite molecules and that the remainder is mostly grossularite. A less abundant kind of garnet is dominantly grossularite mixed with subordinate andradite; this variety of garnet usually has a yellowish hue and is lighter colored than the more common variety. Although this lighter colored garnet is scattered throughout the skarn, it is most common near the contact between skarn and limestone. Most of the garnet is anhedral, but it commonly has crystal faces where it is adjacent to calcite. The pyroxene is greenish gray, and analyses presented by Umpleby (1917, p. 64) show it to be diopside that contains considerable magnesium and a little iron. Locally the skarn consists almost entirely of a mixture of magnetite and hematite, and attempts have been made to exploit a large mass of iron-rich skarn about 1,000 feet northwest of the Grand Prize mine.

To evaluate the changes that occurred during the metasomatism of limestone and intrusive rock to skarn, it is necessary to compare the composition of the rocks before and after metasomatism. Comparison of table 2, which includes analyses of several of the more common types of skarn, with the analyses of limestone and leucogranite porphyry shown in table 1, shows that to form skarn from limestone without change of volume, large amounts of calcium oxide and carbon dioxide must be removed and large amounts of silicon, iron, and aluminum oxides, in decreasing order of abundance, must be added. On the other hand, the principal changes that would be necessary to transform the intrusive rocks to skarn would be the substitution of calcium and a little iron for part of the intrusive rock, principally silica.

The proximity of all the skarn to intrusive rocks suggests that the material which was added to the limestone to form the skarn was related to and may have come from the intrusive rocks or the magma from which they were derived. Furthermore, the fact that all the larger masses of skarn are near leucogranite porphyry suggests that the leucogranite porphyry may have provided more material than the other intrusive rocks. Because all the intrusive rocks seem to be related, a comparison between the composition of the leucogranite porphyry and the other intrusive rocks, especially the Mackay Granite, which is believed to be older than the leucogranite porphyry, might indcate whether or not the leucogranite or the magma from which it was derived has lost any significant amount of material. Inspection of the analyses, table 1, and of the variation diagram, figure 2, does not

TABLE 2.—*Chemical analyses of common types of skarn*

[From Umpleby (1917, p. 63–64). Analysts: Chase Palmer, 7, 8, 6; T. T. Read, 12, 13; Cyril Knight, 9]

	7	12	8	9	6	13
	Normal massive garnet	Massive garnet	Dark-amber garnet, distinct crystals	Light-amber garnet	Massive pyroxene rock	Dark pyroxene rock
SiO_2	36.92	37.79	36.57	37.07	51.55	45.85
Al_2O_3	8.75	11.97	7.56	17.42	4.00	12.21
Fe_2O_3	16.85	15.77	20.34	10.81	1.02	2.15
FeO	.50	1.31	1.24	.68	6.65	2.49
MgO	.17	.37	2.10	.51	11.38	8.70
CaO	33.71	32.57	30.20	32.77	24.33	28.54
Na_2O	.31				.38	
K_2O	.31				.18	
H_2O+	.39		.54	.39	.25	
H_2O-	.21	.09	.30	.14	.14	
TiO_2	.26		.20		.32	
P_2O_5	.30		.23		.24	
MnO	.67	.31	.60		.30	
CO_2	.95					
Total	100.20	100.18	99.88	99.79	100.64	99.94

reveal any systematic impoverishment of the leucogranite porphyry in the three principal elements needed to convert limestone to skarn. It may be significant that the leucogranite porphyry contains somewhat less silicon and iron than the Mackay Granite. Aluminum, which is the third element that is needed to convert limestone to skarn, however, is more abundant in the leucogranite porphyry than in the Mackay Granite.

The texture of the diopside in the leucogranite porphyry also suggests that some of the constituent that makes up the skarn may have come from the leucogranite porphyry. As already noted, diopside is the only mafic mineral that we observed in any appreciable quantity in the leucogranite porphyry, and it occurs as skeletal crystals and aggregates of grains, which seem to be relics of some other mafic minerals, perhaps biotite and hornblende. Alteration of biotite and hornblende to diopside would result in release of aluminum, iron, and perhaps some magnesium, which might then be available to combine with limestone to form skarn. We have found no variation in composition within the leucogranite porphyry however, that could be attributed to differential loss of material from that rock.

Alternatively, the nearness of all the thick skarn to the contact between the leucogranite porphyry and the limestone suggests that this contact may have served as an avenue for movement of fluids that carried material to and from the rocks to form skarn. That this contact is a complex surface that encompasses roof pendants and engulfed blocks of limestone leads us to believe that introduction and removal of material from the rocks along the contact by transient fluids was probably less important than exchange of material across the contact.

JASPEROID

The jasperoid consists of somewhat ferruginous fine-grained quartz, which is commonly brecciated and recemented with fine-grained quartz.

All the larger masses of jasperoid in the area occur along the contact between limestone and the Challis Volcanics, and most of the jasperoid seems to have resulted from replacement of limestone by silica-bearing material.

Most of the jasperoid occurs along faults, which were probably the avenues for the movement of the fluids that caused the jasperoidization. The origin of the silica in these fluids is not clear; it probably came from igneous rocks, but whether these rocks were intrusive or extrusive, and whether the silica was derived from the igneous rocks prior to, during, or even after the rocks had solidified cannot be determined in most places.

Jasperoid a few feet thick is common where the Challis Volcanics has been deposited on limestone, and this jasperoid seems clearly to have been formed by silicification of limestone by fluids from the overlying volcanic rocks. The resulting jasperoid is somewhat more resistant to erosion than the enclosing rocks, and it commonly remains for some time after the overlying volcanic rocks have been eroded away. The jasperoid that is shown on plate 1 in the SE¼ sec. 19, and some of it in sec. 30, T. 7 N., R. 24 E., probably were formed in this manner. Some of the jasperoid elsewhere in sec. 19, T. 7 N., R. 24 E., sec. 31, T. 7 N., R. 24 E., and sec. 24, T. 7 N., R. 23 E., may have also been formed in this way.

SURFICIAL DEPOSITS

GLACIAL DEPOSITS

Patches of boulders and gravel, including fragments of Mackay Granite, occur on two ridges of volcanic rock in the southeastern part of the area These deposits, mapped as older glacial deposits on plate 1, seem to be material that was transported by glaciers or glacial streams that were related to topography that subsequently has been considerably modified.

During the latter part of Pleistocene time the mountains in and near the area of this report were occupied by alpine glaciers, which sculptured the cirques now prominent in the mountains. Deposits of sand, gravel, and silt that are only slightly modified and that are probably of Wisconsin age, designated younger glacial deposits on plate 1, are found in many of the cirques and along many of the streams in the area.

ALLUVIUM

Alluvium of two ages is present in the area shown on plate 1. The older is more extensive and occurs as fans which have been locally dissected by the present streams. The younger is confined to the flood plains along streams.

STRUCTURE

The Paleozoic rocks are rather tightly folded in folds with axes which have a general northwesterly trend. The limestone is locally folded in an intricate and nonsystematic manner, which is characteristic of limestones. Ross (1947, p. 1129–1132) has described these folds in nearby areas, and little discussion is warranted here. Section A–A', plate 1, illustrates some of the larger folds that occur in the Paleozoic rocks.

The Challis Volcanics lie on a major unconformity on the Paleozoic rocks and are folded in gentle nonsystematic undulations.

The geologic map, plate 1, shows two lines which are believed to be due to faulting; both are near boundaries between the Challis Volcanics and Paleozoic rocks, mostly limestone; one trends north-northeastward along the eastern edge of the area, and the other trends northeastward through the northwestern part of the area.

The presence of jasperoid, especially of brecciated jasperoid, along the first of these lines is strong evidence of faulting, although some of the jasperoid along this line may be due to silicification of limestone caused by fluids from volcanic rocks that once overlay the limestone.

The evidence for faulting along the second line consists almost solely of the straightness of the escarpment near the line. It seems likely that this escarpment has retreated southward from the fault, and that some of the uppermost volcanic rocks here have been deposited across the fault subsequent to most, if not all, of the movement on the fault.

Movement on both these faults probably began before or early during the eruption of the Challis Volcanics and may have continued after the volcanic rocks had been extruded.

The dike zone which trends northeastward through the area is evidence of fracturing and perhaps faulting. The fractures in this zone must have developed between the time of emplacement of the Mackay Granite and the time of emplacement of the dikes.

ORE DEPOSITS

Virtually all the mineralization in the Alder Creek mining district is intimately associated with metamorphic rocks, especially skarn, which in turn are confined to contacts between limestone and intrusive rocks. As previously noted, all the thicker masses of skarn seem to be associated with leucogranite porphyry. Nearly 50 properties, mostly on separate lodes, have been worked at one time or another, but most of the production has come from the Empire, Horseshoe, White Knob, Blue Bird, and Champion mines, especially the Empire, which is a copper mine. The Horseshoe and some of the other relatively small mines in the area were developed mainly for their lead-silver content. The map, plate 1, shows the location of the major mines in the district, but it does not adequately indicate the extent to which the hillsides are peppered with prospects, adits, shafts, and pits.

The Empire mine has open-pit and underground workings in a north-trending arcuate zone (Farwell and Full, 1944, p. 12) about 3,500 feet long and as much as 400 feet wide. The horizontal workings have an aggregate length of more than 60,000 feet on more than nine levels, scattered through a vertical range of over 1,500 feet. The lowest exploration is in the 1,600-foot level, or Cossack tunnel, at about 7,010 feet above sea level. The main part of the mine is at the 700-foot level, which is also the longest level. Most of the ore so far mined came from stopes above this level through a vertical range of roughly 500 feet, but ore was mined long ago from stopes extending 300 feet below the 700-foot level.

In 1956 an adit, corresponding approximately to the 1,100-foot level, was driven more than 1,500 feet; during the drilling of this adit, a sulphide ore body was discovered. Exploration has continued intermittenly since then, and in 1960 ore was reported to have been shipped from this adit. In 1961 a mill was constructed at the portal. Sections B-B' and C-C', plate 1, indicate the extent of the workings of the Empire mine. Section B-B' passes lengthwise through the most intensively developed part of the mine, and section C-C' extends across this part of the mine. As previously noted, the control for these cross sections came from detailed maps and cross section of the Empire mine by Farwell and Full. More information on the location of the workings and rocks of the mine can be found in the report of Farwell and Full (1944).

The main haulageways in the Empire mine parallel a segment of the arc within which nearly all the productive mines occur. Most of the ore stopes occur along crosscuts which are approximately radial to the haulageways. These crosscuts follow zones of shearing which cut both the igneous rock and the skarn. These zones, according to Mr. Ray

Webber, who was long the manager of the mine, have proved to be of much value as guides to ore. Many dikes are approximately parallel to this shearing. In most of the stopes visited during an inspection in 1929 (Ross, 1930), the ore was seen to be bounded by fairly definite walls along which there were indications of movement. Some of these walls corresponded to bedding, others to faults.

Valuable concentrations of ore minerals in the Empire mine occur as irregular pipelike bodies (Farwell and Full, 1944, p. 10) that are commonly elliptical in plan, with their long axes in various attitudes but usually steeply plunging. The major axes of the elliptical cross-sections are 15–200 feet long, and the short axes are 5–55 feet long. One pipe has been mined almost continuously through a vertical distance of 600 feet, but most are less persistent. Most pipes pitch northeast, east, or southeast in a direction nearly at right angles to their strikes. Some branch upward; a few downward. A few, mostly at the contact between skarn and marble, are tabular. The ore bodies commonly occur along the contact between skarn and limestone; some are entirely within skarn; and others are at the contact between skarn and instrusive rocks. According to Farwell and Full (1944, p. 13) the ore-bearing skarn is commonly coarser grained and contains more calcite than does the skarn without valuable concentrations of ore minerals. They concluded from this that calcite-rich portions of the skarn were favorable sites for ore deposition. They noted that the concentration of ore minerals diminishes gradually into barren skarn but ends abruptly where the ore-bearing skarn is in contact with limestone.

Chalcopyrite is the principal hypogene mineral in the Empire mine and pyrite, pyrrhotite, calcite, quartz, magnetite, fluorite, scheelite, molybdenite, sphalerite, specularite, and rare bornite are less widely distributed (Farwell and Full, 1944, p. 11). Chrysocolla, accompanied by malachite, tenorite, azurite, and sparse copper sulfate minerals, is the supergene mineral of principal economic importance.

The relative distribution of supergene and hypogene minerals is erratic and not well known. Unoxidized sulfides remain within 50 feet of the surface in a few places, but oxidized ore minerals have been found along sheared zones as deep as the 1,000-foot level although they are more common above the 600-foot level. The Empire mine is developed in a steep mountainside that has over 2,000 feet of local relief, and the mine lies entirely above the foot of the mountain. Throughout this mine, oxidation has occurred wherever the rock has been suitably sheared, but no stable zone seems to have been established wherein secondary enrichment could take place. Only trivial amounts of chalcocite and covellite have been noted.

Umpleby (1917, p. 45–49) gave descriptions and sketches of ore bodies that were accessible to him. The sketches, especially, suggest that the ore bodies are localized and their form influenced by faults and joints. Faults are not shown on Farwell and Full's (1944) surface maps nor on plate 1 of the present report. They are, however, shown on Farwell and Full's underground maps, but everywhere without displacement irrespective of the rocks traversed by the faults. Farwell and Full take cognizance of this fact by noting that displacement along these faults seems to be small. Faults trending northeastward and dipping southeastward are said to persist from level to level, and they are marked by gouge consisting of decomposed intrusive rock, clay, and limonite with some calcite, pyrite, and crushed garnet in zones 2 inches to 6 feet in width (Farwell and Full, 1944, p. 12–14). Some dikes seem to have been intruded after the ore bodies were formed.

Kiilsgaard (in U.S. Geological Survey, 1962) found that in the lower levels of the Empire mine, steeply plunging, irregular, pipelike bodies of primary ore formed where northeast-trending premineralization faults cut the skarn. He reasoned that jointed and fractured skarn was more permeable to mineralizing solutions ascending along the faults than were the limestones and intrusive rocks. Kiilsgaard's conclusions, in part, support the earlier conclusions of Umpleby (1917, p. 45–49) and those of Ross (1930, p. 15, 17), which are less clearly expressed in Ross' brief early paper than in the present report. These conclusions imply that the ore deposition was later than the formation of the skarn, even though there probably is a genetic relationship. This suggests that the ore may persist to a greater depth than might otherwise be possible.

Much of the contact zone between the leucogranite porphyry and limestone has not been located and explored. We suggest that locating this contact zone and searching along it for skarn, followed by systematic exploration of any skarn discovered is likely to reveal additional ore. Farwell and Full (1944, p. 44) made similar recommendations and, in addition, suggested that further exploration of known skarn bodies may locate additional ore and that undiscovered masses of skarn may occur within the leucogranite porphyry wherever it has not been explored.

The mineral deposits on either side of the Empire mine differ in several respects from those in that mine. The Champion mine, at the southern end of the arc delineated by mines, is mostly in limestone, but the workings extend into granitic rock. Some skarn is present in the area, but that reported by Umpleby (1917, p. 101–102) is roughly 1,500 feet from the main workings, which are reached through west-trending adits. There are various small dikes on the property. Apparently most of the lead ore so far found came from one or more crushed or sheared zones in limestone, presumably of westerly trend. The Grand Prize, White Knob, and Blue Bird properties appear to have been worked mainly for lead, but little information about them is on record.

The Horseshoe mine has been explored on six levels, of which the deepest is the 350-foot level. When visited in 1929 (Ross, 1930, p. 15–16), most of the stopes were filled and the lower levels were full of water. Apparently little has been done since that time. The principal workings trend about N. 30° W., and the main ore bodies are along these workings instead of at large angles to them as in the Empire mine. Some of the numerous minor mineralized slips at various angles to the main ore bodies are approximately perpendicular to the general trend; their direction corresponds to the radiate shearing in the Empire mine. Ore deposition has been effected mainly by replacement rather than by fissure fillings, but the shearing that preceded the mineralization was stronger than that found in the Empire mine and was concentrated mainly along the arcuate trend of the district rather than on radial lines. Furthermore, much of the limestone country rock has been recrystallized to marble rather than converted to skarn. Shipments from this mine have been mainly of lead carbonate ore although bodies of sulfide containing pyrite, sphalerite (or more properly marmatite), galena, and chalcopyrite are known. The high iron content of the sphalerite makes it difficult to market. Oxidized material extends below the 200-foot level although there is some sulfide above this level. Most of the production has come from above the 100-foot level.

Several of the lodes in this part of the district contain tungsten (Cook, 1956, p. 27–29). These include the Empire mine, the Phoenix property near the head of Mammoth Canyon in which powellite and scheelite occur in skarn, the Vaught property close to Cliff Creek on which these same two minerals have been found, the Hanni mine near the head of Cliff Creek from which 11 tons of ore assaying 1.60 percent WO_3 were shipped in November 1953, and the Copper Queen workings at the head of the East Fork of Navarro Creek where scheelite has been reported to occur in skarn.

PRODUCTION

Prospecting was carried on in the Alder Creek mining district in the early 1880's, and the first recorded production from the Empire mine was in 1902, although some ore probably was mined earlier. The

Empire mine produced continuously from 1902 through 1930, first under the White Knob Copper Co., and later under the Empire Copper Co., the Idaho Metals Co., the Mackay Metals Co. (Ross, 1930, p. 8–9), the Mackay Exploration Co., and finally, the Lost River Mines Inc. Umpleby (1917, p. 94) estimated the production through 1913 at $2,500,000, mainly in copper, and Ross (1930, p. 8–9) presented production figures from the beginning of 1914 through April 1929 that total over $6,000,000 in large part from leasing operations. Table 3 gives production figures for the period from 1901 through 1942. These records, which were obtained by Farwell and Full (1944, p. 5) from the Metal Economics Division of the Bureau of Mines, are, except for a few years, larger than those which Ross (1930, p. 8–9) obtained from the Mackay Metals Co., and presumably they include production not on record in the company files. Figures on production subsequent to 1942 are not readily available.

Much of the ore shipped from the Empire mine was oxidized material that could not be concentrated by methods then available; this ore is reported to have contained 4–5 percent copper. In addition, concentrates of sulfide ore averaging about half this amount of copper were shipped (Ross, 1930, p. 15). Some early shipments were sulfide ore containing as much as 6 percent copper (Umpleby, 1917, p. 49). At least one carload of tungsten ore containing 2.08 percent WO_3 was shipped from below the 1,000-foot level of the Empire in 1942 (Farwell and Full, 1944, p. 1, 11). During the most active part of the history of the Empire, only fairly rich copper ore or material easily concentrated by gravity methods could be profitably mined, even though a mill and smelter were built at that time. Presumably much copper-bearing material that could be profitably mined and processed, using more modern techniques, remains in the mine. Some may be already mined and on the dumps.

Production from the Horseshoe mine from 1916 through 1928, as presented by Farwell and Full (1944, p. 6), is summarized in table 4. The mine closed in 1930.

The Champion mine, discovered in 1895, is estimated (J. L. Ausich, written commun., 1961) to have yielded about 4,500 tons of ore through 1960.

Other mines and prospects in the vicinity were in operation until about 1930.

TABLE 3.—*Production of recovered metals at the Empire mine, 1901–42*

These statistics have been furnished by the Metal Economics Division of the Bureau of Mines. Comparison with company records indicate that the crude ore figures include only ore shipped or smelted and not crude ore concentrated]

Year	Crude ore (dry tons)	Concentrates (dry tons)	Gold (ounces)	Silver (ounces)	Copper (pounds)
1901	None				
1902	1,721		14.40	607	14,966
1903	15,681		240.95	12,658	441,286
1904	67,850		85.	3,500	2,700,000
1905	13,000		384.74	22,065	684,134
1906	40,838		1,842.	71,854	2,807,926
1907	37,141	3,430	1,823.33	70,222	2,895,881
1908	382		15.89	673	38,698
1909	1,436		27.73	2,236	90,347
1910	7,206		265.24	28,754	919,492
1911	11,057		663.	40,900	1,415,314
1912	26,227		1,766.	69,942	2,854,281
1913	35,950		1,891.61	106,463	3,962,125
1914	17,801		970.99	59,243	2,106,441
1915	54,295		3,155.06	125,134	4,702,119
1916	69,907		2,874.60	123,453	5,006,291
1917	66,808		2,530.	74,645	4,208,401
1918	53,211		2,476.41	56,014	3,404,161
1919	12,904		672.80	31,833	1,300,518
1920	15,755		1,369.	29,888	1,480,678
1921	9,992		1,236.	23,354	1,088,148
1922	16,717		2,019.	33,988	1,843,200
1923	15,791		1,458.	25,908	1,449,838
1924	11,775	319	1,244.92	18,808	1,137,771
1925	29,753	4,760	2,096.43	35,439	2,352,306
1926	3,635	255	234.38	6,453	239,785
1927	13,627	1,297	761.22	9,734	684,154
1928	11,532	1,053	495.	9,776	514,697
1929	66,573	4,273	2,282.45	60,883	2,824,032
1930	26,214	2,379	754.51	22,925	1,121,586
1931	None				
1932	None				
1933	None				
1934	None				
1935	190		10.10	1,510	26,518
1936	173		8.83	639	18,897
1937	22		1.00	306	3,876
1938	None				
1939	996		207.	2,465	175,940
1940	4,484		526.	11,300	632,217
1941	3,169		381.	7,013	380,469
1942	1,274		141.	1,874	104,000
Total [1]	765,087	17,766	36,925.59	1,202,459	55,630,493

[1] In addition a small tonnage of tungsten ore has been produced.

TABLE 4.—*Production of the Horseshoe mine, 1916–28*

[From records of the U.S. Bureau of Mines]

Year	Crude ore (tons)	Concentrates (tons)	Gold (ounces)	Silver (ounces)	Copper (pounds)	Lead (pounds)	Zinc (pounds)
1916	196		0.88	1,468	3,093	55,587	
1917	1,462		6.99	17,293	2,997	678,074	
1918	1,087		8.64	11,440	1,522	446,628	
1919	1,128		4.76	12,011	2,333	361,937	
1920	2,319		5.96	21,165	2,021	386,352	
1921	116		2.25	1,811	376	55,902	
1922	651		7.61	9,297	902	246,659	
1923	660		6.98	5,014	1,603	169,872	
1924	79		4.33	1,012	287	36,538	
1925	75		1.50	1,776	348	59,310	
1926	865	32	11.72	10,236	3,198	332,597	18,942
1927	182	53	1.52	2,275	495	58,256	18,163
1928	49	5	.43	123	1,159	791	2,366
Total	8,869	90	63.57	94,917	20,334	2,888,503	39,471

REFERENCES CITED

Cook, E. F., 1956, Tungsten deposits of south-central Idaho: Idaho Bur. Mines and Geology Pamph. 108, 40 p.

Farwell, F. W., and Full, R. P., 1944, Geology of the Empire Copper mine near Mackay, Idaho: U.S. Geol. Survey open-file report, October 1944, 22 p.

Kemp, J. F., and Gunther, C. G., 1908, The White Knob copper-deposits, Mackay, Idaho: Am. Inst. Mining Metall. Engineers Trans., v. 38, p. 269–296

Ross, C. P., 1930, The Alder Creek Mining district, Custer County, Idaho *in* Geology and ore deposits of the Seafoam, Alder Creek, Little Smoky, and Willow Creek mining districts, Custer and Camas Counties, Idaho; Idaho Bur. Mines and Geology Pamph. 33, p. 7–18.

———— 1937, Geology and ore deposits of the Bayhorse region, Custer County, Idaho: U.S. Geol. Survey Bull. 877, 161 p.

———— 1947, Geology of the Borah Peak quadrangle, Idaho; Geol. Soc. America Bull., v. 58, no. 12, pt. 1, p. 1085–1160.

———— 1960, Diverse interfingering Carboniferous strata in the Mackay quadrangle, Idaho *in* Geological Survey research 1960: U.S. Geol. Survey Prof. Paper 400–B, p. B232–B233.

———— 1961, A redefinition and restriction of the term Challis volcanics, *in* Geological Survey research 1961: U.S. Geol. Survey Prof. Paper 424–C, p. C177–C180.

———— 1962, Upper Paleozoic rocks in central Idaho: Am. Assoc. Petroleum Geologists Bull., v. 46, no. 3, p. 384–387.

Skipp, B. A. L., 1961, Interpretation of sedimentary features in Brazer Limestone (Mississippian) near Mackay, Custer County, Idaho: Am. Assoc. Petroleum Geologists Bull., v. 45, no. 3, p. 376–389.

Umpleby, J. B., 1917, Geology and ore deposits of the Mackay region, Idaho: U.S. Geol. Survey Prof. Paper 97, 129 p.

U.S. Geological Survey, 1962, Copper ore in Idaho, *in* Geological Survey research 1962: U.S. Geol. Survey Prof. Paper 450–A, p. A3–A4.

○

GOLD RUSH BOOKS

OREGON, USA

www.GoldMiningBooks.com

Books On Mining

Visit: www.goldminingbooks.com to order your copies or ask your favorite book seller to offer them.

Mining Books by Kerby Jackson

Gold Dust: Stories From Oregon's Mining Years - Oregon mining historian and prospector, Kerby Jackson, brings you a treasure trove of seventeen stories on Southern Oregon's rich history of gold prospecting, the prospectors and their discoveries, and the breathtaking areas they settled in and made homes. 5" X 8", 98 ppgs. Retail Price: $11.99

The Golden Trail: More Stories From Oregon's Mining Years - In his follow-up to "Gold Dust: Stories of Oregon's Mining Years", this time around, Jackson brings us twelve tales from Oregon's Gold Rush, including the story about the first gold strike on Canyon Creek in Grant County, about the old timers who found gold by the pail full at the Victor Mine near Galice, how Iradel Bray discovered a rich ledge of gold on the Coquille River during the height of the Rogue River War, a tale of two elderly miners on the hunt for a lost mine in the Cascade Mountains, details about the discovery of the famous Armstrong Nugget and others. 5" X 8", 70 ppgs. Retail Price: $10.99

Oregon Mining Books

Geology and Mineral Resources of Josephine County, Oregon - Unavailable since the 1970's, this important publication was originally compiled by the Oregon Department of Geology and Mineral Industries and includes important details on the economic geology and mineral resources of this important mining area in South Western Oregon. Included are notes on the history, geology and development of important mines, as well as insights into the mining of gold, copper, nickel, limestone, chromium and other minerals found in large quantities in Josephine County, Oregon. 8.5" X 11", 54 ppgs. Retail Price: $9.99

Mines and Prospects of the Mount Reuben Mining District - Unavailable since 1947, this important publication was originally compiled by geologist Elton Youngberg of the Oregon Department of Geology and Mineral Industries and includes detailed descriptions, histories and the geology of the Mount Reuben Mining District in Josephine County, Oregon. Included are notes on the history, geology, development and assay statistics, as well as underground maps of all the major mines and prospects in the vicinity of this much neglected mining district. 8.5" X 11", 48 ppgs. Retail Price: $9.99

The Granite Mining District - Notes on the history, geology and development of important mines in the well known Granite Mining District which is located in Grant County, Oregon. Some of the mines discussed include the Ajax, Blue Ribbon, Buffalo, Continental, Cougar-Independence, Magnolia, New York, Standard and the Tillicum. Also included are many rare maps pertaining to the mines in the area. 8.5" X 11", 48 ppgs. Retail Price: $9.99

Ore Deposits of the Takilma and Waldo Mining Districts of Josephine County, Oregon - The Waldo and Takilma mining districts are most notable for the fact that the earliest large scale mining of placer gold and copper in Oregon took place in these two areas. Included are details about some of the earliest large gold mines in the state such as the Llano de Oro, High Gravel, Cameron, Platerica, Deep Gravel and others, as well as copper mines such as the famous Queen of Bronze mine, the Waldo, Lily and Cowboy mines. This volume also includes six maps and 20 original illustrations. 8.5" X 11", 74 ppgs. Retail Price: $9.99

Metal Mines of Douglas, Coos and Curry Counties, Oregon - Oregon mining historian Kerby Jackson introduces us to a classic work on Oregon's mining history in this important re-issue of Bulletin 14C Volume 1, otherwise known as the Douglas, Coos & Curry Counties, Oregon Metal Mines Handbook. Unavailable since 1940, this important publication was originally compiled by the Oregon Department of Geology and Mineral Industries includes detailed descriptions, histories and the geology of over 250 metallic mineral mines and prospects in this rugged area of South West Oregon. 8.5" X 11", 158 ppgs. Retail Price: $19.99

Metal Mines of Jackson County, Oregon - Unavailable since 1943, this important publication was originally compiled by the Oregon Department of Geology and Mineral Industries includes detailed descriptions, histories and the geology of over 450 metallic mineral mines and prospects in Jackson County, Oregon. Included are such famous gold mining areas as Gold Hill, Jacksonville, Sterling and the Upper Applegate. **8.5" X 11", 220 ppgs. Retail Price: $24.99**

Metal Mines of Josephine County, Oregon - Oregon mining historian Kerby Jackson introduces us to a classic work on Oregon's mining history in this important re-issue of Bulletin 14C, otherwise known as the Josephine County, Oregon Metal Mines Handbook. Unavailable since 1952, this important publication was originally compiled by the Oregon Department of Geology and Mineral Industries includes detailed descriptions, histories and the geology of over 500 metallic mineral mines and prospects in Josephine County, Oregon. **8.5" X 11", 250 ppgs. Retail Price: $24.99**

Metal Mines of North East Oregon - Oregon mining historian Kerby Jackson introduces us to a classic work on Oregon's mining history in this important re-issue of Bulletin 14A and 14B, otherwise known as the North East Oregon Metal Mines Handbook. Unavailable since 1941, this important publication was originally compiled by the Oregon Department of Geology and Mineral Industries and includes detailed descriptions, histories and the geology of over 750 metallic mineral mines and prospects in North Eastern Oregon. **8.5" X 11", 310 ppgs. Retail Price: $29.99**

Metal Mines of North West Oregon - Oregon mining historian Kerby Jackson introduces us to a classic work on Oregon's mining history in this important re-issue of Bulletin 14D, otherwise known as the North West Oregon Metal Mines Handbook. Unavailable since 1951, this important publication was originally compiled by the Oregon Department of Geology and Mineral Industries and includes detailed descriptions, histories and the geology of over 250 metallic mineral mines and prospects in North Western Oregon. **8.5" X 11", 182 ppgs. Retail Price: $19.99**

Mines and Prospects of Oregon - Mining historian Kerby Jackson introduces us to a classic mining work by the Oregon Bureau of Mines in this important re-issue of The Handbook of Mines and Prospects of Oregon. Unavailable since 1916, this publication includes important insights into hundreds of gold, silver, copper, coal, limestone and other mines that operated in the State of Oregon around the turn of the 19th Century. Included are not only geological details on early mines throughout Oregon, but also insights into their history, production, locations and in some cases, also included are rare maps of their underground workings. **8.5" X 11", 314 ppgs. Retail Price: $24.99**

Lode Gold of the Klamath Mountains of Northern California and South West Oregon
(See California Mining Books)

Mineral Resources of South West Oregon - Unavailable since 1914, this publication includes important insights into dozens of mines that once operated in South West Oregon, including the famous gold fields of Josephine and Jackson Counties, as well as the Coal Mines of Coos County. Included are not only geological details on early mines throughout South West Oregon, but also insights into their history, production and locations. **8.5" X 11", 154 ppgs. Retail Price: $11.99**

Chromite Mining in The Klamath Mountains of California and Oregon
(See California Mining Books)

Southern Oregon Mineral Wealth - Unavailable since 1904, this rare publication provides a unique snapshot into the mines that were operating in the area at the time. Included are not only geological details on early mines throughout South West Oregon, but also insights into their history, production and locations. Some of the mining areas include Grave Creek, Greenback, Wolf Creek, Jump Off Joe Creek, Granite Hill, Galice, Mount Reuben, Gold Hill, Galls Creek, Kane Creek, Sardine Creek, Birdseye Creek, Evans Creek, Foots Creek, Jacksonville, Ashland, the Applegate River, Waldo, Kerby and the Illinois River, Althouse and Sucker Creek, as well as insights into local copper mining and other topics. **8.5" X 11", 64 ppgs. Retail Price: $8.99**

Geology and Ore Deposits of the Takilma and Waldo Mining Districts - Unavailable since the 1933, this publication was originally compiled by the United States Geological Survey and includes details on gold and copper mining in the Takilma and Waldo Districts of Josephine County, Oregon. The Waldo and Takilma mining districts are most notable for the fact that the earliest large scale mining of placer gold and copper in Oregon took place in these two areas. Included in this report are details about some of the earliest large gold mines in the state such as the Llano de Oro, High Gravel, Cameron, Platerica, Deep Gravel and others, as well as copper mines such as the famous Queen of Bronze mine, the Waldo, Lily and Cowboy mines. In addition to geological examinations, insights are also provided into the production, day to day operations and early histories of these mines, as well as calculations of known mineral reserves in the area. This volume also includes six maps and 20 original illustrations. **8.5" X 11", 74 ppgs. Retail Price: $9.99**

Gold Mines of Oregon - Oregon mining historian Kerby Jackson introduces us to a classic work on Oregon's mining history in this important re-issue of Bulletin 61, otherwise known as "Gold and Silver In Oregon". Unavailable since 1968, this important publication was originally compiled by geologists Howard C. Brooks and Len Ramp of the Oregon Department of Geology and Mineral Industries and includes detailed descriptions, histories and the geology of over 450 gold mines Oregon. Included are notes on the history, geology and gold production statistics of all the major mining areas in Oregon including the Klamath Mountains, the Blue Mountains and the North Cascades. While gold is where you find it, as every miner knows, the path to success is to prospect for gold where it was previously found. 8.5" X 11", 344 ppgs. **Retail Price: $24.99**

Mines and Mineral Resources of Curry County Oregon - Originally published in 1916, this important publication on Oregon Mining has not been available for nearly a century. Included are rare insights into the history, production and locations of dozens of gold mines in Curry County, Oregon, as well as detailed information on important Oregon mining districts in that area such as those at Agness, Bald Face Creek, Mule Creek, Boulder Creek, China Diggings, Collier Creek, Elk River, Gold Beach, Rock Creek, Sixes River and elsewhere. Particular attention is especially paid to the famous beach gold deposits of this portion of the Oregon Coast. 8.5" X 11", 140 ppgs. **Retail Price: $11.99**

Chromite Mining in South West Oregon - Originally published in 1961, this important publication on Oregon Mining has not been available for nearly a century. Included are rare insights into the history, production and locations of nearly 300 chromite mines in South Western Oregon. 8.5" X 11", 184 ppgs. **Retail Price: $14.99**

Mineral Resources of Douglas County Oregon - Originally published in 1972, this important publication on Oregon Mining has not been available for nearly forty years. Included are rare insights into the geology, history, production and locations of numerous gold mines and other mining properties in Douglas County, Oregon. 8.5" X 11", 124 ppgs. **Retail Price: $11.99**

Mineral Resources of Coos County Oregon - Originally published in 1972, this important publication on Oregon Mining has not been available for nearly forty years. Included are rare insights into the geology, history, production and locations of numerous gold mines and other mining properties in Coos County, Oregon. 8.5" X 11", 100 ppgs. **Retail Price: $11.99**

Mineral Resources of Lane County Oregon - Originally published in 1938, this important publication on Oregon Mining has not been available for nearly seventy five years. Included are extremely rare insights into the geology and mines of Lane County, Oregon, in particular in the Bohemia, Blue River, Oakridge, Black Butte and Winberry Mining Districts. 8.5" X 11", 82 ppgs. **Retail Price: $9.99**

Mineral Resources of the Upper Chetco River of Oregon: Including the Kalmiopsis Wilderness - Originally published in 1975, this important publication on Oregon Mining has not been available for nearly forty years. Withdrawn under the 1872 Mining Act since 1984, real insight into the minerals resources and mines of the Upper Chetco River has long been unavailable due to the remoteness of the area. Despite this, the decades of battle between property owners and environmental extremists over the last private mining inholding in the area has continued to pique the interest of those interested in mining and other forms of natural resource use. Gold mining began in the area in the 1850's and has a rich history in this geographic area, even if the facts surrounding it are little known. Included are twenty two rare photographs, as well as insights into the Becca and Morning Mine, the Emmly Mine (also known as Emily Camp), the Frazier Mine, the Golden Dream or Higgins Mine, Hustis Mine, Peck Mine and others. 8.5" X 11", 64 ppgs. **Retail Price: $8.99**

Gold Dredging in Oregon - Originally published in 1939, this important publication on Oregon Mining has not been available for nearly seventy five years. Included are extremely rare insights into the history and day to day operations of the dragline and bucketline gold dredges that once worked the placer gold fields of South West and North East Oregon in decades gone by. Also included are details into the areas that were worked by gold dredges in Josephine, Jackson, Baker and Grant counties, as well as the economic factors that impacted this mining method. This volume also offers a unique look into the values of river bottom land in relation to both farming and mining, in how farm lands were mined, re-soiled and reclamated after the dredges worked them. Featured are hard to find maps of the gold dredge fields, as well as rare photographs from a bygone era. 8.5" X 11", 86 ppgs. **Retail Price: $8.99**

Quick Silver Mining in Oregon - Originally published in 1963, this important publication on Oregon Mining has not been available for over fifty years. This publication includes details into the history and production of Elemental Mercury or Quicksilver in the State of Oregon. 8.5" X 11", 238 ppgs. **Retail Price: $15.99**

Mines of the Greenhorn Mining District of Grant County Oregon - Originally published in 1948, this important publication on Oregon Mining has not been available for over sixty five years. In this publication are rare insights into the mines of the famous Greenhorn Mining District of Grant County, Oregon, especially the famous Morning Mine. Also included are details on the Tempest, Tiger, Bi-Metallic, Windsor, Psyche, Big Johnny, Snow Creek, Banzette and Paramount Mines, as well as prospects in the vicinities in the famous mining areas of Mormon Basin, Vinegar Basin and Desolation Creek. Included are hard to find mine maps and dozens of rare photographs from the bygone era of Grant County's rich mining history. 8.5" X 11", 72 ppgs. **Retail Price: $9.99**

Geology of the Wallowa Mountains of Oregon: Part I (Volume 1) - Originally published in 1938, this important publication on Oregon Mining has not been available for nearly seventy five years. Included are details on the geology of this unique portion of North Eastern Oregon. This is the first part of a two book series on the area. Accompanying the text are rare photographs and historic maps.8.5" X 11", 92 ppgs. Retail Price: $9.99

Geology of the Wallowa Mountains of Oregon: Part II (Volume 2) - Originally published in 1938, this important publication on Oregon Mining has not been available for nearly seventy five years. Included are details on the geology of this unique portion of North Eastern Oregon. This is the first part of a two book series on the area. Accompanying the text are rare photographs and historic maps.8.5" X 11", 94 ppgs. Retail Price: $9.99

Field Identification of Minerals For Oregon Prospectors - Originally published in 1940, this important publication on Oregon Mining has not been available for nearly seventy five years. Included in this volume is an easy system for testing and identifying a wide range of minerals that might be found by prospectors, geologists and rockhounds in the State of Oregon, as well as in other locales. Topics include how to put together your own field testing kit and how to conduct rudimentary tests in the field. This volume is written in a clear and concise way to make it useful even for beginners. 8.5" X 11", 158 ppgs. Retail Price: $14.99

The Bohemia Mining District of Oregon - Originally published in 1900, this important publication on Oregon Mining has not been available for over a century. Included in this volume are important insights into the famous Bohemia Mining District of Oregon, including the histories and locations of important gold mines in the area such as the Ophir Mine, Clarence, Acturas, Peek-a-boo, White Swan, Combination Mine, the Musick Mine, The California, White Ghost, The Mystery, Wall Street, Vesuvius, Story, Lizzie Bullock, Delta, Elsie Dora, Golden Slipper, Broadway, Champion Mine, Knott, Noonday, Helena, White Wings, Riverside and others. Also included are notes on the nearby Blue River Mining District. 8.5" X 11", 58 ppgs. Retail Price: $9.99

The Gold Fields of Eastern Oregon - Unavailable since 1900, this publication was originally compiled by the Baker City Chamber of Commerce Offering important insights into the gold mining history of Eastern Oregon, "The Gold Fields of Eastern Oregon" sheds a rare light on many of the gold mines that were operating at the turn of the 19th Century in Baker County and Grant County in North Eastern Oregon. Some of the areas featured include the Cable Cove District, Baisely-Elhorn, Granite, Red Boy, Bonanza, Susanville, Sparta, Virtue, Vaughn, Sumpter, Burnt River, Rye Valley and other mining districts. Included is basic information on not only many gold mines that are well known to those interested in Eastern Oregon mining history, but also many mines and prospects which have been mostly lost to the passage of time. Accompanying are numerous rare photos 8.5" X 11", 78 ppgs. Retail Price: $10.99

Gold Mining in Eastern Oregon - Originally published in 1938, this important publication on Oregon Mining has not been available for over a century. Included in this volume are important insights into the famous mining districts of Eastern Oregon during the late 1930's. Particular attention is given to those gold mines with milling and concentrating facilities in the Greenhorn, Red Boy, Alamo, Bonanza, Granite, Cable Cove, Cracker Creek, Virtue, Keating, Medical Springs, Sanger, Sparta, Chicken Creek, Mormon Basin, Connor Creek, Cornucopia and the Bull Run Mining Districts. Some of the mines featured include the Ben Harrison, North Pole-Columbia, Highland Maxwell, Baisley-Elkhorn, White Swan, Balm Creek, Twin Baby, Gem of Sparta, New Deal, Gleason, Gifford-Johnson, Cornucopia, Record, Bull Run, Orion and others. Of particular interest are the mill flow sheets and descriptions of milling operations of these mines. 8.5" X 11", 68 ppgs. Retail Price: $8.99

The Gold Belt of the Blue Mountains of Oregon - Originally published in 1901, this important publication on Oregon Mining has not been available for over a century. Included in this volume are rare insights into the gold deposits of the Blue Mountains of North East Oregon, including the history of their early discovery and early production. Extensive details are offered on this important mining area's mineralogy and economic geology, as well as insights into nearby gold placers, silver deposits and copper deposits. Featured are the Elkhorn and Rock Creek mining districts, the Pocahontas district, Auburn and Minersville districts, Sumpter and Cracker Creek, Cable Cove, the Camp Carson district, Granite, Alamo, Greenhorn, Robinsonville, the Upper Burnt River Valley and Bonanza districts, Susanville, Quartzburg, Canyon Creek, Virtue, the Copper Butte district, the North Powder River, Sparta, Eagle Creek, Cornucopia, Pine Creek, Lower Powder River, the Upper Snake River Canyon, Rye Valley, Lower Burnt River Valley, Mormon Basin, the Malheur and Clarks Creek districts, Sutton Creek and others. Of particular interest are important details on numerous gold mines and prospects in these mining districts, including their locations, histories, geology and other important information, as well as information on silver, copper and fire opal deposits. 8.5" X 11", 250 ppgs. Retail Price: $24.99

Mining in the Cascades Range of Oregon - Originally published in 1938, this important publication on Oregon Mining has not been available for over seventy five years. Included in this volume are rare insights into the gold mines and other types of metal mines in the Cascades Mountain Range of Oregon. Some of the important mining areas covered include the famous Bohemia Mining District, the North Santiam Mining District, Quartzville Mining District, Blue River Mining District, Fall Creek Mining District, Oakridge District, Zinc District, Buzzard-Al Sarena District, Grand Cove, Climax District and Barron Mining District. Of particular interest are important details on over 100 mines and prospects in these mining districts, including their locations, histories, geology and other important information. 8.5" X 11", 170 ppgs. Retail Price: $14.99

Beach Gold Placers of the Oregon Coast - Originally published in 1934, this important publication on Oregon Mining has not been available for over 80 years. Included in this volume are rare insights into the beach gold deposits of the State of Oregon, including their locations, occurance, composition and geology. Of particular interest is information on placer platinum in Oregon's rich beach deposits. Also included are the locations and other information on some famous Oregon beach mines, including the Pioneer, Eagle, Chickamin, Iowa and beach placer mines north of the mouth of the Rogue River. 8.5" X 11", 60 ppgs. Retail Price: $8.99

Mineralogical Composition of the Sands of the Oregon Coast: From Coos Bay to the Columbia - Published in 1945, he text features hard to find information on the composition of the gold bearing black sands of the South West Oregon Coast, offering a unique insight to prospectors in search of Oregon's legendary beach gold. 104 ppgs, $9.99

Manganese Mining in Oregon - First released in 1942 and now out of print, this special reprint edition of "Manganese in Oregon" was originally published by the Oregon Department of Geology and Mineral Industries. The text features hard to find information on the mining of Manganese in Oregon, including details and maps of Oregon manganese mines and prospects. 108 ppgs, 9.99

Medford Oregon As A Mining Center - Written in 1912, this hard to find publication includes valuable insights into the mining history of South West Oregon. This small book contains interesting information on the gold, copper and mining industry in Southern Oregon as it existed just prior to World War One, shedding light on some of the important mines in the area. Included are rare photographs and vintage advertising of the day. 80 ppgs, 9.99

Mineral Resources of Curry County Oregon - First released in 1977 and now out of print, this special reprint edition of "Geology, Mineral Resources and Rock Materials of Curry County, Oregon" was originally published in cooperation of Curry County, Oregon and the Oregon Department of Geology and Mineral Industries. The text features hard to find information on not only the mining of gold and other metals in Curry County, but also aggregate mining in the area. 102 ppgs, 11.99

Origin of the Gold Bearing Black Sands of the Coast of South West Oregon - First released in 1943 and now out of print, this special reprint edition of "The Origin of the Black Sands of the South West Oregon Coast" was originally published by the Oregon Department of Geology and Mineral Industries. The text features hard to find information on the origin of the gold bearing black sands of the South West Oregon Coast, offering a unique insight to prospectors in search of Oregon's legendary beach gold. 52 ppgs, 8.99

South West Oregon Mining - Leading mining historian Kerby Jackson introduces us to six classic small mining publications on the Gold Mining Industry in Southern Oregon. This small book consists of a compilation of USGS J.S. Diller's "Mines of the Riddles Quadrangle", "The Rogue River Valley Coal Fields" and "Mineral Resources of the Grants Pass Quadrangle", the Grants Pass Commercial Club's rare publication "Mining in Josephine County, Oregon" and the USGS publication "The Distribution of Placer Gold in the Sixes River, South West Oregon". Also included is F.W. Libbey's legendary article on the Southern Oregon Mining Industry, "Lest We Forget", which appeared in the publication of the Oregon State Department of Geology and Mineral Industries in the early 1960's. This compilation offers a unique perspective on mining in South West Oregon and includes considerable information on mines in Josephine, Jackson and Coos Counties. 142 ppgs, 14.99

Geology and Mineral Resources of the Gasquet Quadrangle of California-Oregon - First published in 1953, it has been unavailable for over a century and sheds important light on the geological features and mineral resources of this portion of Northern California and Southern Oregon. 80 ppgs, 9.99

Idaho Mining Books

Gold in Idaho - Unavailable since the 1940's, this publication was originally compiled by the Idaho Bureau of Mines and includes details on gold mining in Idaho. Included is not only raw data on gold production in Idaho, but also valuable insight into where gold may be found in Idaho, as well as practical information on the gold bearing rocks and other geological features that will assist those looking for placer and lode gold in the State of Idaho. This volume also includes thirteen gold maps that greatly enhance the practical usability of the information contained in this small book detailing where to find gold in Idaho. **8.5" X 11", 72 ppgs. Retail Price: $9.99**

Geology of the Couer D'Alene Mining District of Idaho - Unavailable since 1961, this publication was originally compiled by the Idaho Bureau of Mines and Geology and includes details on the mining of gold, silver and other minerals in the famous Coeur D'Alene Mining District in Northern Idaho. Included are details on the early history of the Coeur D'Alene Mining District, local tectonic settings, ore deposit features, information on the mineral belts of the Osburn Fault, as well as detailed information on the famous Bunker Hill Mine, the Dayrock Mine, Galena Mine, Lucky Friday Mine and the infamous Sunshine Mine. This volume also includes sixteen hard to find maps. **8.5" X 11", 70 ppgs. Retail Price: $9.99**

The Gold Camps and Silver Cities of Idaho - Originally published in 1963, this important publication on Idaho Mining has not been available for nearly fifty years. Included are rare insights into the history of Idaho's Gold Rush, as well as the mad craze for silver in the Idaho Panhandle. Documented in fine detail are the early mining excitements at Boise Basin, at South Boise, in the Owyhees, at Deadwood, Long Valley, Stanley Basin and Robinson Bar, at Atlanta, on the famous Boise River, Volcano, Little Smokey, Banner, Boise Ridge, Hailey, Leesburg, Lemhi, Pearl, at South Mountain, Shoup and Ulysses, Yellow Jacket and Loon Creek. The story follows with the appearance of Chinese miners at the new mining camps on the Snake River, Black Pine, Yankee Fork, Bay Horse, Clayton, Heath, Seven Devils, Gibbonsville, Vienna and Sawtooth City. Also included are special sections on the Idaho Lead and Silver mines of the late 1800's, as well as the mining discoveries of the early 1900's that paved the way for Idaho's modern mining and mineral industry. Lavishly illustrated with rare historic photos, this volume provides a one of a kind documentary into Idaho's mining history that is sure to be enjoyed by not only modern miners and prospectors who still scour the hills in search of nature's treasures, but also those enjoy history and tromping through overgrown ghost towns and long abandoned mining camps. **8.5" X 11", 186 ppgs. Retail Price: $14.99**

Ore Deposits and Mining in North Western Custer County Idaho - Unavailable since 1913, this important publication was originally published by the Us Department of the Interior and has been unavailable for a century. Included are fine details on the geology, geography, gold placers and gold and silver bearing quartz veins of the mining region of North West Custer County, Idaho. Of particular interest is a rare look at the mines and prospects of the region, including those such as the Ramshorn Mine, SkyLark, Riverview, Excelsior, Beardsley, Pacific, Hoosier, Silver Brick, Forest Rose and dozens of others in the Bay Horse Mining District. Also covered are the mines of the Yankee Fork District such as the Lucky Boy, Badger, Black, Enterprise, Charles Dickens, Morrison, Golden Sunbeam, Montana, Golden Gate and others, as well as those in the Loon Mining District. **8.5" X 11", 126 ppgs. Retail Price: $12.99**

Gold Rush To Idaho - Unavailable since 1963, this important publication was originally published by the Idaho Bureau of Mines and has been unavailable for 50 years. "Gold Rush To Idaho" revisits the earliest years of the discovery of gold in Idaho Territory and introduces us to the conditions that the pioneer gold seekers met when they blazed a trail through the wilderness of Idaho's mountains and discovered the precious yellow metal at Oro Fino and Pierce. Subsequent rushes followed at places like Elk City, Newsome, Clearwater Station, Florence, Warrens and elsewhere. Of particular interest is a rare look at the hardships that the first miners in Idaho met with during their day to day existences and their attempts to bring law and order to their mining camps. **8.5" X 11", 88 ppgs. Retail Price: $9.99**

The Geology and Mines of Northern Idaho and North Western Montana - Unavailable since 1909, this important publication was originally published by the Us Department of the Interior and has been unavailable for a century. Included are fine details on the geology and geography of the mining regions of Northern Idaho and North Western Montana. Of particular interest is a rare look at the mines and prospects of the region, including those in the Pine Creek Mining District, Lake Pend Oreille district, Troy Mining District, Sylvanite District, Cabinet Mining District, Prospect Mining District and the Missoula Valley. Some of the mines featured include the Iron Mountain, Silver Butte, Snowshoe, Grouse Mountain Mine and others. **8.5" X 11", 142 ppgs. Retail Price: $12.99**

Mining in the Alturas Quadrangle of Blaine County Idaho - Unavailable since 1922, this important publication was originally published by the Idaho Bureau of Mines and has been unavailable for ninety years. Topics include the geology, rock formations and the formation of ore deposits in this important mining area of Idaho. Of particular focus is information on the local geology, quartz veins and ore deposits of this portion of Idaho. Included are hard to find details, including the descriptions and locations of numerous gold and silver mines in the area including the Silver King, Pilgrim, Columbia, Lone Jack, Sunbeam, Pride of the West, Lucky Boy, Scotia, Atlanta, Beaver-Bidwell and others mines and prospects. **8.5" X 11", 56 ppgs. Retail Price: $8.99**

Mining in Lemhi County Idaho - Originally published in 1913, this important book on Idaho Mining has not been available to miners for over a century. Included are rare insights into hundreds of gold, silver, copper and other mines in this famous Idaho mining area. Details include the locations, geology, history, production and other facts of the mines of this region, not only gold and silver hardrock mines, but also gold placer mines, lead-silver deposits, copper mines, cobalt-nickel deposits, tungsten and tin mines . It is lavishly illustrated with hard to find photos of the period and rare mining maps. Some of the vicinities featured include the Nicholia Mining District, Spring Mountain District, Texas District, Blue Wing District, Junction District, McDevitt District, Pratt Creek, Eldorado District, Kirtley Creek, Carmen Creek, Gibbonsville, Indian Creek, Mineral Hill District, Mackinaw, Eureka District, Blackbird District, YellowJacket District, Gravel Range District, Junction District, Parker Mountain and other mining districts. 8.5" X 11", 226 ppgs. **Retail Price: $19.99**

Mining in Shoshone County Idaho - First published in 1923, it has been unavailable for over a century and sheds important light on the mining history of Shoshone County, Idaho. Some of the topics include the history of mining in Shoshone County, a look at the local geology and ore characteristics of lead-silver deposits, zinc deposits, copper, antimony, gold and other minerals. Also included are insights into the history, production, characteristics and locations of numerous mines in the area. 198 ppgs, 15.99

Utah Mining Books

Fluorite in Utah - Unavailable since 1954, this publication was originally compiled by the USGS, State of Utah and U.S. Atomic Energy Commission and details the mining of fluorspar, also known as fluorite in the State of Utah. Included are details on the geology and history of fluorspar (fluorite) mining in Utah, including details on where this unique gem mineral may be found in the State of Utah. 8.5" X 11", 60 ppgs. **Retail Price: $8.99**

The Gold Hill Mining District of Utah - First published in 1935, it has been unavailable since those days and sheds important light on the mines, history and geology of Utah's Gold Hill Mining District. Included are rare insights into this important mining area, including the locations, histories and details of numerous mines. This volume is well illustrated with geological diagrams, as well as hard to find maps of some of the most important mines in this district. 202 ppgs., 19.99

The Mines, Miners and Minerals of Utah - First published in 1896, it has been unavailable since those days and sheds important light on the early mines and miners of Pioneer Utah, as well as the minerals which they won from the earth by laborious hard physical labor and sheer determination. Included are rare insights into the early mining history of Utah, as well details on hundreds of gold, silver and copper mines. 376 ppgs., 24.99

California Mining Books

The Tertiary Gravels of the Sierra Nevada of California - Mining historian Kerby Jackson introduces us to a classic mining work by Waldemar Lindgren in this important re-issue of The Tertiary Gravels of the Sierra Nevada of California. Unavailable since 1911, this publication includes details on the gold bearing ancient river channels of the famous Sierra Nevada region of California. 8.5" X 11", 282 ppgs. **Retail Price: $19.99**

The Mother Lode Mining Region of California - Unavailable since 1900, this publication includes details on the gold mines of California's famous Mother Lode gold mining area. Included are details on the geology, history and important gold mines of the region, as well as insights into historic mining methods, mine timbering, mining machinery, mining bell signals and other details on how these mines operated. Also included are insights into the gold mines of the California Mother Lode that were in operation during the first sixty years of California's mining history. 8.5" X 11", 176 ppgs. Retail Price: $14.99

Lode Gold of the Klamath Mountains of Northern California and South West Oregon - Unavailable since 1971, this publication was originally compiled by Preston E. Hotz and includes details on the lode mining districts of Oregon and California's Klamath Mountains. Included are details on the geology, history and important lode mines of the French Gulch, Deadwood, Whiskeytown, Shasta, Redding, Muletown, South Fork, Old Diggings, Dog Creek (Delta), Bully Choop (Indian Creek), Harrison Gulch, Hayfork, Minersville, Trinity Center, Canyon Creek, East Fork, New River, Denny, Liberty (Black Bear), Cecilville, Callahan, Yreka, Fort Jones and Happy Camp mining districts in California, as well as the Ashland, Rogue River, Applegate, Illinois River, Takilma, Greenback, Galice, Silver Peak, Myrtle Creek and Mule Creek districts of South Western Oregon. Also included are insights into the mineralization and other characteristics of this important mining region. 8.5" X 11", 100 ppgs. **Retail Price: $10.99**

Mines and Mineral Resources of Shasta County, Siskiyou County, Trinity County: California - Unavailable since 1915, this publication was originally compiled by the California State Mining Bureau and includes details on the gold mines of this area of Northern California. Also included are insights into the mineralization and other characteristics of this important mining region, as well as the location of historic gold mines. 8.5" X 11", 204 ppgs. **Retail Price: $19.99**

Geology of the Yreka Quadrangle, Siskiyou County, California - Unavailable since 1977, this publication was originally compiled by Preston E. Hotz and includes details on the geology of the Yreka Quadrangle of Siskiyou County, California. Also included are insights into the mineralization and other characteristics of this important mining region. **8.5" X 11", 78 ppgs. Retail Price: $7.99**

Mines of San Diego and Imperial Counties, California - Originally published in 1914, this important publication on California Mining has not been available for a century. This publication includes important information on the early gold mines of San Diego and Imperial County, which were some of the first gold fields mined in California by early Spanish and Mexican miners before the 49ers came on the scene. Included are not only details on early mining methods in the area, production statistics and geological information, but also the location of the early gold mines that helped make California "The Golden State". Also included are details on the mining of other minerals such as silver, lead, zinc, manganese, tungsten, vanadium, asbestos, barite, borax, cement, clay, dolomite, fluospar, gem stones, graphite, marble, salines, petroleum, stronium, talc and others. **8.5" X 11", 116 ppgs. Retail Price: $12.99**

Mines of Sierra County, California - Unavailable since 1920, this publication was originally compiled by the California State Mining Bureau and includes details on the gold mines of Sierra County, California. Also included are insights into the mineralization and other characteristics of this important mining region, as well as the location of historic gold mines. **8.5" X 11", 156 ppgs. Retail Price: $19.99**

Mines of Plumas County, California - Unavailable since 1918, this publication was originally compiled by the California State Mining Bureau and includes details on the gold mines of Plumas County, California. Also included are insights into the mineralization and other characteristics of this important mining region, as well as the location of historic gold mines. **8.5" X 11", 200 ppgs. Retail Price: $19.99**

Mines of El Dorado, Placer, Sacramento and Yuba Counties, California - Originally published in 1917, this important publication on California Mining has not been available for nearly a century. This publication includes important information on the early gold mines of El Dorado County, Placer County, Sacramento County and Yuba County, which were some of the first gold fields mined by the Forty-Niners during the California Gold Rush. Included are not only details on early mining methods in the area, production statistics and geological information, but also the location of the early gold mines that helped make California "The Golden State". Also included are insights into the early mining of chrome, copper and other minerals in this important mining area. **8.5" X 11", 204 ppgs. Retail Price: $19.99**

Mines of Los Angeles, Orange and Riverside Counties, California - Originally published in 1917, this important publication on California Mining has not been available for nearly a century. This publication includes important information on the early gold mines of Los Angeles County, Orange County and Riverside County, which were some of the first gold fields mined in California by early Spanish and Mexican miners before the 49ers came on the scene. Included are not only details on early mining methods in the area, production statistics and geological information, but also the location of the early gold mines that helped make California "The Golden State". **8.5" X 11", 146 ppgs. Retail Price: $12.99**

Mines of San Bernadino and Tulare Counties, California - Originally published in 1917, this important publication on California Mining has not been available for nearly a century. This publication includes important information on the early gold mines of San Bernadino and Tulare County, which were some of the first gold fields mined in California by early Spanish and Mexican miners before the 49ers came on the scene. Included are not only details on early mining methods in the area, production statistics and geological information, but also the location of the early gold mines that helped make California "The Golden State". Also included are details on the mining of other minerals such as copper, iron, lead, zinc, manganese, tungsten, vanadium, asbestos, barite, borax, cement, clay, dolomite, fluospar, gem stones, graphite, marble, salines, petroleum, stronium, talc and others. **8.5" X 11", 200 ppgs. Retail Price: $19.99**

Chromite Mining in The Klamath Mountains of California and Oregon - Unavailable since 1919, this publication was originally compiled by J.S. Diller of the United States Department of Geological Survey and includes details on the chromite mines of this area of Northern California and Southern Oregon. Also included are insights into the mineralization and other characteristics of this important mining region, as well as the location of historic mines. Also included are insights into chromite mining in Eastern Oregon and Montana. **8.5" X 11", 98 ppgs. Retail Price: $9.99**

Mines and Mining in Amador, Calaveras and Tuolumne Counties, California - Unavailable since 1915, this publication was originally compiled by William Tucker and includes details on the mines and mineral resources of this important California mining area. Included are details on the geology, history and important gold mines of the region, as well as insights into other local mineral resources such as asbestos, clay, copper, talc, limestone and others. Also included are insights into the mineralization and other characteristics of this important portion of California's Mother Lode mining region. **8.5" X 11", 198 ppgs. Retail Price: $14.99**

The Cerro Gordo Mining District of Inyo County California - Unavailable since 1963, this publication was originally compiled by the United States Department of Interior. Included are insights into the mineralization and other characteristics of this important mining region of Southern California. Topics include the mining of gold and silver in this important mining district in Inyo County, California, including details on the history, production and locations of the Cerro Gordo Mine, the Morning Star Mine, Estelle Tunnel, Charles Lease Tunnel, Ignacio, Hart, Crosscut Tunnel, Sunset, Upper Newtown, Newtown, Ella, Perseverance, Newsboy, Belmont and other silver and gold mines in the Cerro Gordo Mining District. This volume also includes important insights into the fossil record, geologic formations, faults and other aspects of economic geology in this California mining district. 8.5" X 11", 104 ppgs. Retail Price: $10.99

Mining in Butte, Lassen, Modoc, Sutter and Tehama Counties of California - Unavailable since 1917, this publication was originally compiled by the United States Department of Interior. Included are insights into the mineralization and other characteristics of this important mining region of California. Topics include the mining of asbestos, chromite, gold, diamonds and manganese in Butte County, the mining of gold and copper in the Hayden Hill and Diamond Mountain mining districts of Lassen County, the mining of coal, salt, copper and gold in the High Grade and Winters mining districts of Modoc County, gold mining in Sutter County and the mining of gold, chromite, manganese and copper in Tehama County. This volume also includes the production records and locations of numerous mines in this important mining region. 8.5" X 11", 114 ppgs. Retail Price: $11.99

Mines of Trinity County California - Originally published in 1965, this important publication on California Mining has not been available for nearly fifty years. This publication includes important information on mines and mining in Trinity County, California, as well insights into the mineralization and geology of this important mining area in Northern California. Included are extensive details on hardrock and placer gold mines and prospects, including charts showing the locations of these historic mines.. 8.5" X 11", 144 ppgs. Retail Price: $12.99

Mines of Kern County California - Originally published in 1962, this important publication on California Mining has not been available for nearly fifty years. This publication includes important information on mines and mining in Kern County, California, as well insights into the mineralization and geology of this important mining area in California. Included are extensive details on hardrock and placer gold mines and prospects, including charts showing the locations of these historic mines. 8.5" X 11", 398 ppgs. Retail Price: $24.99

Mines of Calaveras County California - Originally published in 1962, this important publication on California Mining has not been available for nearly fifty years. This publication includes important information on mines and mining in Calaveras County, California, as well insights into the mineralization and geology of this important mining area in Northern California. Included are extensive details on hardrock and placer gold mines and prospects, including charts showing the locations of these historic mines. 8.5" X 11", 236 ppgs. Retail Price: $19.99

Lode Gold Mining in Grass Valley California - Unavailable since 1940, this publication was originally compiled by the United States Department of Interior. Included are insights into the gold mineralization and other characteristics of this important mining region of Nevada County, California. This volume also includes important insights into the geologic formations, faults and other aspects of economic geology in this California mining district. Of particular interest are the fine details on many hardrock gold mines in the area, including their locations, histories, development and mineralization. Some of the mines featured include the Gold Hill Mine, Massachusetts Hill, Boundary, Peabody, Golden Center, North Star, Omaha, Lone Jack, Homeward Bound, Hartery, Wisconsin, Allison Ranch, Phoenix, Kate Hayes, W.Y.O.D., Empire, Rich Hill, Daisy Hill, Orleans, Sultana, Centennial, Conlin, Ben Franklin, Crown Point and many others. 8.5" X 11", 148 ppgs. Retail Price: $12.99

Lode Mining in the Alleghany District of Sierra County California - Unavailable since 1913, this publication was originally compiled by the United States Department of Interior. Included are insights into the mineralization and other characteristics of this important mining region of Sierra County. Included are details on the history, production and locations of numerous hardrock gold mines in this famous California area, including the Tightner Mine, Minnie D., Osceola, Eldorado, Twenty One, Sherman, Kenton, Oriental, Rainbow, Plumbago, Irelan, Gold Canyon, North Fork, Federal, Kate Hardy and others. This volume also includes important insights into the fossil record, geologic formations, faults and other aspects of economic geology in this California mining district. 8.5" X 11", 48 ppgs. Retail Price: $7.99

Six Months In The Gold Mines During The California Gold Rush - Unavailable since 1850, this important work is a first hand account of one "49'ers" personal experience during the great California Gold Rush, shedding important light on one of the most exciting periods in the history of not only California, but also the world. Compiled from journals written between 1847 and 1849 by E. Gould Buffum, a native of New York, "Six Months In The Gold Mines During The California Gold Rush" offers a rare look into the day to day lives of the people who came to California to work in her gold mines when the state was still a great frontier. 8.5" X 11", 290 ppgs. Retail Price: $19.99

<u>Quartz Mines of the Grass Valley Mining District of California</u> - Unavailable since 1867, this important publication has not been available since those days. This rare publication offers a short dissertation on the early hardrock mines in this important mining district in the California Mother Lode region between the 1850's and 1860's. Also included are hard to find details on the mineralization and locations of these mines, as well as how they were operated in those day. **8.5" X 11", 44 ppgs. Retail Price: $8.99**

<u>Gold Rush on the Feather River</u> - **First published in 1924, this short publication by G.C. Mansfield sheds important light on the early history of gold mining on the Feather River. Included are rare insights into the first decade of gold mining and the early mining camps of the Feather River during the 1850's. 64 ppgs., 9.99**

<u>The Bodie Mining District of California</u> - **First published in 1986, it has been unavailable since those days and sheds important light on this famous mining area. Included are the history, characteristics and locations of numerous old mines around the ghost town of Bodie.**
64 ppgs, 8.99

<u>Geology and Mineral Resources of the Gasquet Quadrangle of California-Oregon</u> - **First published in 1953, it has been unavailable for over a century and sheds important light on the geological features and mineral resources of this portion of Northern California and Southern Oregon.**
80 ppgs, 9.99

Alaska Mining Books

<u>Ore Deposits of the Willow Creek Mining District, Alaska</u> - Unavailable since 1954, this hard to find publication includes valuable insights into the Willow Creek Mining District near Hatcher Pass in Alaska. The publication includes insights into the history, geology and locations of the well known mines in the area, including the Gold Cord, Independence, Fern, Mabel, Lonesome, Snowbird, Schroff-O'Neil, High Grade, Marion Twin, Thorpe, Webfoot, Kelly-Willow, Lane, Holland and others. **8.5" X 11", 96 ppgs. Retail Price: $9.99**

<u>The Juneau Gold Belt of Alaska</u> - Unavailable since 1906, this hard to find publication includes valuable insights into the gold mines around Juneau, Alaska. The publication includes important details into the history, geology and locations of the well known gold mines and prospects in the area, including those around Windham Bay, Holkham Bay, Port Snettisham, on Grindstone and Rhine Creeks, Gold Creek, Douglas Island, Salmon Creek, Lemon Creek, Nugget Creek, from the Mendenhall River to Berners Bay, McGinnis Creek, Montana Creek, Peterson Creek, Windfall Creek, the Eagle River, Yankee Basin, Yankee Curve, Kowee Creek and elsewhere. Not only are gold placer mines included, but also hardrock gold mines. **8.5" X 11", 224 ppgs. Retail Price: $19.99**

<u>Mining in the Jumbo Basin of Alaska</u> - **Unavailable since 1953, this hard to find publication includes valuable insights into the mines and geology of the Jumbo Basin. The publication includes important details into the history, geology and locations of the well known gold mines and prospects in the famous Jumbo Basin Mining Region of Alaska.**
72 ppgs, 9.99

<u>The Rampart Placer Gold Region of Alaska</u> - Unavailable since 1906, this hard to find publication includes valuable insights into the placer gold mines of the Rampart Mining Region. The publication includes important details into the history, geology and locations of the well known gold mines and prospects in the famous Rampart Mining Region of Alaska.
78 ppgs, 10.99

Arizona Mining Books

<u>Mines and Mining in Northern Yuma County Arizona</u> - Originally published in 1911, this important publication on Arizona Mining has not been available for over a hundred years. Included are rare insights into the gold, silver, copper and quicksilver mines of Yuma County, Arizona together with hard to find maps and photographs. Some of the mines and mining districts featured include the Planet Copper Mine, Mineral Hill, the Clara Consolidated Mine, Viati Mine, Copper Basin prospect, Bowman Mine, Quartz King, Billy Mack, Carnation, the Wardwell and Osbourne, Valensuella Copper, the Mariquita, Colonial Mine, the French American, the New York-Plomosa, Guadalupe, Lead Camp, Mudersbach Copper Camp, Yellow Bird, the Arizona Northern (Salome Strike), Bonanza (Harqua Hala), Golden Eagle, Hercules, Socorro and others. **8.5" X 11", 144 ppgs. Retail Price: $11.99**

<u>The Aravaipa and Stanley Mining Districts of Graham County Arizona</u> - Originally published in 1925, this important publication on Arizona Mining has not been available for nearly ninety years. Included are rare insights into the gold and silver mines of these two important mining districts, together with hard to find maps. **8.5" X 11", 140 ppgs. Retail Price: $11.99**

Gold in the Gold Basin and Lost Basin Mining Districts of Mohave County, Arizona - This volume contains rare insights into the geology and gold mineralization of the Gold Basin and Lost Basin Mining Districts of Mohave County, Arizona that will be of benefit to miners and prospectors. Also included is a significant body of information on the gold mines and prospects of this portion of Arizona. This volume is lavishly illustrated with rare photos and mining maps. **8.5" X 11", 188 ppgs. Retail Price: $19.99**

Mines of the Jerome and Bradshaw Mountains of Arizona - This important publication on Arizona Mining has not been available for ninety years. This volume contains rare insights into the geology and ore deposits of the Jerome and Bradshaw Mountains of Arizona that will be of benefit to miners and prospectors who work those areas. Included is a significant body of information on the mines and prospects of the Verde, Black Hills, Cherry Creek, Prescott, Walker, Groom Creek, Hassayampa, Bigbug, Turkey Creek, Agua Fria, Black Canyon, Peck, Tiger, Pine Grove, Bradshaw, Tintop, Humbug and Castle Creek Mining Districts. This volume is lavishly illustrated with rare photos and mining maps. **8.5" X 11", 218 ppgs. Retail Price: $19.99**

The Ajo Mining District of Pima County Arizona - This important publication on Arizona Mining has not been available for nearly seventy years. This volume contains rare insights into the geology and mineralization of the Ajo Mining District in Pima County, Arizona and in particular the famous New Cornelia Mine. **8.5" X 11", 126 ppgs. Retail Price: $11.99**

Mining in the Santa Rita and Patagonia Mountains of Arizona - Originally published in 1915, this important publication on Arizona Mining has not been available for nearly a century. Included are rare insights into hundreds of gold, silver, copper and other mines in this famous Arizona mining area. Details include the locations, geology, history, production and other facts of the mines of this region. **8.5" X 11", 394 ppgs. Retail Price: $24.99**

Mining in the Bisbee Quadrangle of Arizona - Originally published in 1906, this important publication on Arizona Mining has not been available for nearly a century. Included are rare insights into hundreds of gold, silver, copper and other mines in this famous Arizona mining area. Details include the locations, geology, history, production and other facts of the mines of this important mining region. **8.5" X 11", 188 ppgs. Retail Price: $14.99**

Placer Gold Mining in Arizona - Unavailable since 1922, this hard to find publication includes valuable insights into the placer gold mines of the Arizona. Originally released as "Placer Gold of Arizona", despite its small size, this publication includes important details into the history, geology and locations of the well known placer gold mines and prospects in the State of Arizona. 48 ppgs, 8.99

Gold and Copper Mining near Payson, Arizona - Written in 1915, this hard to find publication includes valuable insights into the gold and copper mining industry of Arizona. Highlighted here are the gold and copper mines near Payson, Arizona. 68 ppgs, 8.99

Lode Gold Mining in Arizona - Unavailable since 1934, this hard to find publication, originally released as "Arizona Lode Gold Mines and Gold Mining" includes valuable insights into the gold mining industry of Arizona. Included are valuable insights into over 150 hardrock gold mines in over 30 different mining districts in Arizona. 278 ppgs, 21.99

Mining in the Dragoon Quadrangle of Cochise County, Arizona - Unavailable since 1964, this hard to find publication includes valuable insights into the mines of the Dragoon Quadrangle Mining Region. The publication includes important details into the history, geology and locations of the well known mines and prospects in this famous mining region of Arizona. 224 ppgs., 19.99

Directory of Operating Mines in Arizona in 1915 - Unavailable since 1916, this hard to find publication includes valuable insights into the mines of Arizona. This small publication includes a complete list of the mines that were operating in the State of Arizona during 1915 and includes details such as general location, owners and some basic facts about each mining operation. 52 ppgs. 8.99

Arizona Ore Deposits - Unavailable since 1938, this hard to find publication includes valuable insights into some ore deposits of Arizona. Included are valuable insights into the formation and characteristics of valuable ore deposits in the Jerome, Miami, Inspiration, Clifton, Morenci, Ray, Ajo, Eureka, Tombstone and Magma mining districts. Included are details into some of the major gold, silver and copper mines of these important Arizona mining areas. 160 ppgs, 14.99

Montana Mining Books

A History of Butte Montana: The World's Greatest Mining Camp - First published in 1900 by H.C. Freeman, this important publication sheds a bright light on one of the most important mining areas in the history of The West. Together with his insights, as well as rare photographs of the periods, Harry Freeman describes Butte and its vicinity from its early beginnings, right up to its flush years when copper flowed from its mines like a river. At the time of publication, Butte, Montana was known worldwide as "The Richest Mining Spot On Earth" and produced not only vast amounts of copper, but also silver, gold and other metals from its mines. Freeman illustrates, with great detail, the most important mines in the vicinity of Butte, providing rare details on their owners, their history and most importantly, how the mines operated and how their treasures were extracted. Of particular interest are the dozens of rare photographs that depict mines such as the famous Anaconda, the Silver Bow, the Smoke House, Moose, Paulin, Buffalo, Little Minah, the Mountain Consolidated, West Greyrock, Cora, the Green Mountain, Diamond, Bell, Parnell, the Neversweat, Nipper, Original and many others.   8.5" X 11", 142 ppgs. Retail Price: $12.99

The Butte Mining District of Montana - This important publication on Montana Mining has not been available for over a century. Included are rare insights into the gold, copper and silver mines of Butte, Montana together with hard to find maps and photographs. Some of the topics include the early history of gold, silver and copper mining in the Butte area, insight into the geology of its mining areas, the local distribution of gold, silver and copper ores, as well their composition and how to identify them. Also included are detailed facts about the mines in the Butte Mining District, including the famous Anaconda Mine, Gagnon, Parrot, Blue Vein, Moscow, Poulin, Stella, Buffalo, Green Mountain, Wake Up Jim, the Diamond-Bell Group, Mountain Consolidated, East Greyrock, West Greyrock, Snowball, Corra, Speculator, Adirondack, Miners Union, the Jessie-Edith May Group, Otisco, Iduna, Colorado, Lizzie, Cambers, Anderson, Hesperus, Preferencia and dozens of others. 8.5" X 11", 298 ppgs. Retail Price: $24.99

Mines of the Helena Mining Region of Montana - This important publication on Montana Mining has not been available for over a century. Included are rare insights into the gold, copper and silver mines of the vicinity of Helena, Montana, including the Marysville Mining District, Elliston Mining District, Rimini Mining District, Helena Mining District, Clancy Mining District, Wickes Mining District, Boulder and Basin Mining Districts and the Elkhorn Mining District. Some of the topics include the early history of gold, silver and copper mining in the Helena area, insight into the geology of its mining areas, the local distribution of gold, silver and copper ores, as well their composition and how to identify them. Also included are detailed facts, history, geology and locations of over one hundred gold, silver and copper mines in the area . 8.5" X 11", 162 ppgs, Retail Price: $14.99

Mines and Geology of the Garnet Range of Montana - This important publication on Montana Mining has not been available for over a century. Included are rare insights into the gold, copper and silver mines of the vicinity of this important mining area of Montana. Some of the topics include the early history of gold, silver and copper mining in the Garnet Mountains, insight into the geology of its mining areas, the local distribution of gold, silver and copper ores, as well their composition and how to identify them. Also included are detailed facts, history, geology and locations of numerous gold, silver and copper mines in the area . 8.5" X 11", 100 ppgs, Retail Price: $11.99

Mines and Geology of the Philipsburg Quadrangle of Montana - This important publication on Montana Mining has not been available for over a century. Included are rare insights into the gold, copper and silver mines of the vicinity of this important mining area of Montana. Some of the topics include the early history of gold, silver and copper mining in the Philipsburg Quadrangle, insight into the geology of its mining areas, the local distribution of gold, silver and copper ores, as well their composition and how to identify them. Also included are detailed facts, history, geology and locations of over one hundred gold, silver and copper mines in the area 8.5" X 11", 290 ppgs, Retail Price: $24.99

Geology of the Marysville Mining District of Montana - Included are rare insights into the mining geology of the Marysville Mining District. Some of the topics include the early history of gold, silver and copper mining in the area, insight into the geology of its mining areas, the local distribution of gold, silver and copper ores, as well their composition and how to identify them. Also included are detailed facts, history, geology and locations of gold, silver and copper mines in the area 8.5" X 11", 198 ppgs, Retail Price: $19.99

The Geology and Mines of Northern Idaho and North Western Montana- See listing under Idaho.

The History of Gold Dredging in Montana - Unavailable since 1916, this important publication was originally published by the Us Bureau of Mines and has been unavailable for a century. A century and more ago, giant dredging machines dug in Montana's rivers and creeks in search of illusive golden riches. First appearing in California in the 1850's, gold dredges finally reached their peak of development in Siberia and New Zealand before becoming popular again in the United States. This book offers a unique historical perspective on the gold dredges that once operated in Montana. This book on Montana mining history is lavishly illustrated with dozens of rare historic photos gold dredges that once operated in Montana, as well as hard to locate plans on how these dredges were designed. 120 ppgs., 11.99

Nevada Mining Books

The Bull Frog Mining District of Nevada - Unavailable since 1910, this publication was originally compiled by the United States Department of Interior. This volume also includes important insights into the geologic formations, faults and other aspects of economic geology in this Nevada mining district. Of particular interest are the fine details on many mines in the area, including their locations, histories, development and mineralization. Some of the mines featured include the National Bank Mine, Providence, Gibraltor, Tramps, Denver, Original Bullfrog, Gold Bar, Mayflower, Homestake-King and other mines and prospects. **8.5" X 11", 152 ppgs, Retail Price: $14.99**

History of the Comstock Lode - Unavailable since 1876, this publication was originally released by John Wiley & Sons. This volume also includes important insights into the famous Comstock Lode of Nevada that represented the first major silver discovery in the United States. During its spectacular run, the Comstock produced over 192 million ounces of silver and 8.2 million ounces of gold. Not only did the Comstock result in one of the largest mining rushes in history and yield immense fortunes for its owners, but it made important contributions to the development of the State of Nevada, as well as neighboring California. Included here are important details on not only the early development and history of the Comstock, but also rare early insight into its mines, ore and its geology. **8.5" X 11", 244 ppgs, Retail Price: $19.99**

The Pioche Mining District of Nevada - First published in 1932, it has been unavailable for over a century and sheds important light on the mining history of Nevada. Some of the topics include the history of mining in this district, as well as the characteristics of its mineral and ore deposits. Also included are insights into the history, production, characteristics and locations of numerous mines in the area. Some of the mines include the Combined Metals, Pioche, Ely Valley, No. 10, Poorman, Wide Awake, Alps, Prince, Virginia Louise, Half Moon, Abe Lincoln, Fairview, Bristol Silver, National, Vesuvius, Inman, Tempest, Hillside, Jackrabbit, Lucky Star, Fortuna, Mendha, Manhattan, Hamburg, Comet, Lyndon and others. 108 ppgs 10.99

The Yerington Mining District of Nevada - First published in 1932, it has been unavailable for over a century and sheds important light on the mining history of Nevada. Some of the topics include the history of mining in this district, as well as the characteristics of its mineral and ore deposits. Also included are insights into the history, production, characteristics and locations of numerous mines in the area. Some of the mines include the Bluestone, Mason Valley, Malachite, McConnell, Greenwood, Western Nevada, Ludwig, Douglas Hill, Casting Copper, Montana-Yerington, Empire, Jim Beatty, Terry and McFarland, Blue Jay and others. 92 ppgs, 10.99

The Genesis of the Ores of Tonopah Nevada - Unavailable since 1918, this hard to find publication includes valuable insights into the gold mines around Tonopah, Nevada. The publication includes important details into the geology of mines in the Tonopah Mining District of Nevada. 90 ppgs, 10.99

Mining Camps of Elko, Lander and Eureka Counties Nevada - Unavailable since 1910, this hard to find publication includes valuable insights into the mining camps of Elko, Lander and Eureka Counties, Nevada. The publication includes important details into the history of mines and mining in these three Nevada counties. 154 ppgs, 12.99

Ore Deposits of the Bullfrog Quadrangle - Unavailable since 1964 and released as "Geology of Bullfrog Quadrangle and Ore Deposits Related to Bullfrog Hills Caldera, Nye County, Nevada and Inyo County, California". The publication includes important details into the geology of mines in the Bullfrog Quadrangle of Nye County, Nevada and Inyo County, California. 52 ppgs, 9.99

Mining in Eureka County Nevada - Unavailable since 1879, this hard to find publication includes valuable insights into the early mining history off Eureka County, Nevada. The publication includes important details into the early history of the mines of Eureka County, as well as their development, production and how their ores were treated. Also included are details on the 1872 Mining Act, as well as the local rules, regulations and customs of the miners in Eureka County. 134 ppgs, 12.99

Colorado Mining Books

Ores of The Leadville Mining District - Unavailable since 1926, this publication was originally compiled by the United States Department of Interior. This volume also includes important insights into the ores and mineralization of the Leadville Mining District in Colorado. Topics include historic ore prospecting methods, local geology, insights into ore veins and stockworks, the local trend and distribution of ore channels, reverse faults, shattered rock above replacement ore bodies, mineral enrichment in oxidized and sulphide zones and more. **8.5" X 11", 66 ppgs, Retail Price: $8.99**

Mining in Colorado - Unavailable since 1926, this publication was originally compiled by the United States Department of Interior. This volume also includes important insights into the mining history of Colorado from its early beginnings in the 1850's right up to the mid 1920's. Not only is Colorado's gold mining heritage included, but also its silver, copper, lead and zinc mining industry. Each mining area is treated separately, detailing the development of Colorado's mines on a county by county basis. **8.5" X 11", 284 ppgs, Retail Price: $19.99**

Gold Mining in Gilpin County Colorado - Unavailable since 1876, this publication was originally compiled by the Register Steam Printing House of Central City, Colorado. A rare glimpse at the gold mining history and early mines of Gilpin County, Colorado from their first discovery in the 1850's up to the "flush years" of the mid 1870's. Of particular interest is the history of the discovery of gold in Gilpin County and details about the men who made those first strikes. Special focus is given to the early gold mines and first mining districts of the area, many of which are not detailed in other books on Colorado's gold mining history. **8.5" X 11", 156 ppgs, Retail Price: $12.99**

Mining in the Gold Brick Mining District of Colorado - Important insights into the history of the Gold Brick Mining District, as well as its local geography and economic geology. Also included are the histories and locations of historic mines in this important Colorado Mining District, including the Cortland, Carter, Raymond, Gold Links, Sacramento, Bassick, Sandy Hook, Chronicle, Grand Prize, Chloride, Granite Mountain, Lucille, Gray Mountain, Hilltop, Maggie Mitchell, Silver Islet, Revenue, Roosevelt, Carbonate King and others. In addition to hardrock mining, are also included are details on gold placer mining in this portion of Colorado. **8.5" X 11", 140 ppgs, Retail Price: $12.99**

Ore Deposits of the London Fault of Colorado - First published in 1941, it has been unavailable since those days and sheds important light on the mines and mineral deposits of the London Fault in Central Colorado's Alma Mining District. This publication sheds important light on the gold veins and lead-silver deposits of the Alma Mining District. Included are geologic details on the London Mine, American Mine, Havigorst Tunnel, Ophir Mine, Mosher Tunnel, London-Butte Mine, Venture Shaft, Hard-To-Beat Mine, Oliver Twist Tunnel, Sacramento Mine, Mudsill Mine, Sherwood Mine, Wagner, Barcoe Tunnel and other mines in this important mining region. 110 ppgs., 10.99

The Mines of Colorado - First published in 1867, it has been unavailable since those days and sheds important light on Colorado's early mining history. Written shortly after the events took place, this publication sheds important light on the Pike's Peak Gold Rush, the discovery of gold on Ralston Creek and Dry Creek in the 1850's, as well as details on the first wave of miners into Colorado and their trials and tribulations as they crossed the Great Plains. Also included are details on early discoveries of lode gold in the mountainous regions of Colorado, details on the early mines hardrock and placer mines, and much more. It is a veritable treasure trove on Colorado's early mining history and will be of great importance to anyone who is interested in the mining of gold or other minerals in Colorado, as well as those interested in the history of the state. 478 ppgs., 29.99

The La Plata Mining District of Colorado - Originally titled "Geology and Ore Deposits in the Vicinity of the La Plata District of Colorado" and first published in 1949, it has been unavailable since those days and sheds important light on the mines and mineral deposits of the La Plata Mining District of Colorado. 214 ppgs., 19.99

Washington Mining Books

The Republic Mining District of Washington - Unavailable since 1910, this important publication was originally published by the Washington Geologic Survey and has been unavailable for a century. Topics include the geology, rock formations and the formation of ore deposits in this important mining area of Washington State. Also included are hard to find details on the geology, history and locations of dozens of mines in the area. Some of the mines featured include the New Republic Mine, Ben Hur, Morning Glory, the South Republic Mine, Quilp, Surprise, Black Tail, Lone Pine, San Poil, Mountain Lion, Tom Thumb, Elcaliph and many others. **8.5" X 11", 94 ppgs, Retail Price: $10.99**

The Myers Creek and Nighthawk Mining Districts of Washington - Unavailable since 1911, this important publication was originally published by the Washington Geologic Survey and has been unavailable for a century. Topics include the geology, rock formations and the formation of ore deposits in these important mining areas of Washington State. Also included are hard to find details on the geology, history and locations of dozens of mines in the area. Some of the mines featured include the Grant Mine, Monterey, Nip and Tuck, Myers Creek, Number Nine, Neutral, Rainbow, Aztec, Crystal Butte, Apex, Butcher Boy, Molson, Mad River, Olentangy, Delate, Kelsey, Golden Chariot, Okanogan, Ohio, Forty-Ninth Parallel, Nighthawk, Favorite, Little Chopaka, Summit, Number One, California, Peerless, Caaba, Prize Group, Ruby, Mountain Sheep, Golden Zone, Rich Bar, Similkameen, Kimberly, Triune, Hiawatha, Trinity, Hornsilver, Maquae, Bellevue, Bullfrog, Palmer Lake, Ivanhoe, Copper World and many others. **8.5" X 11", 136 ppgs, Retail Price: $12.99**

The Blewett Mining District of Washington - Unavailable since 1911, this important publication was originally published by the Washington Geologic Survey and has been unavailable for a century. Topics include the geology, rock formations and the formation of ore deposits in this important mining area of Washington State. Also included are hard to find details on the geology, history and locations of dozens of mines in the area. Some of the mines featured include the Washington Meteor, Alta Vista, Pole Pick, Blinn, North Star, Golden Eagle, Tip Top, Wilder, Golden Guinea, Lucky Queen, Blue Bell, Prospect, Homestake, Lone Rock, Johnson, and others. **8.5" X 11", 134 ppgs, Retail Price: $12.99**

Silver Mining In Washington - Unavailable since 1955, this important publication was originally published by the Washington Geologic Survey. Featured are the hard to find locations and details pertaining to Washington's silver mines. **8.5" X 11", 180 ppgs, Retail Price: $15.99**

The Mines of Snohomish County Washington - Unavailable since 1942, this important publication was originally published by the Washington Geologic Survey and has been unavailable for seventy years. Featured are details on a large number of gold, silver, copper, lead and other metallic mineral mines. Included are the locations of each historic mine, along with information on the commodity produced. **8.5" X 11", 98 ppgs, Retail Price: $10.99**

The Mines of Chelan County Washington - Unavailable since 1943, this important publication was originally published by the Washington Geologic Survey and has been unavailable for seventy years. Featured are details on a large number of gold, silver, copper, lead and other metallic mineral mines. Included are the locations of each historic mine, along with information on the commodity. **8.5" X 11", 88 ppgs, Retail Price: $9.99**

Metal Mines of Washington - Unavailable since 1921, this important publication was originally published by the Washington Geologic Survey and has been unavailable for nearly ninety years. Widely considered a masterpiece on the Washington Mining Industry, "Metal Mines of Washington" sheds light on the important details of Washington's early mining years. Featured are details on hundreds of gold, silver, copper, lead and other metallic mineral mines. Included are hard to find details on the mineral resources of this state, as well as the locations of historic mines. Lavishly illustrated with maps and historic photos and complete with a glossary to explain any technical terms found in the text, this is one of the most important works on mining in the State of Washington. No prospector or miner should be without it if they are interested in mining in Washington. **8.5" X 11", 396 ppgs, Retail Price: $24.99**

Gem Stones In Washington - Unavailable since 1949, this important publication was originally published by the Washington Geologic Survey and has been unavailable since first published. Included are details on where to find naturally occurring gem stones in the State of Washington, including quartz crystal, amethyst, smoky quartz, milky quartz, agates, bloodstone, carnelian, chert, flint, jasper, onyx, petrified wood, opal, fire opal, hyalite and others. **8.5" X 11", 54 ppgs, Retail Price: $8.99**

The Covada Mining District of Washington - Unavailable since 1913, this important publication was originally published by the Washington Geologic Survey and has been unavailable for a century. Topics include the geology, rock formations and the formation of ore deposits in this important mining area of Washington State. Also included are hard to find details on the geology, history and locations of dozens of mines in the area. Some of the mines featured include the Admiral, Advance, Algonkian, Big Bug, Big Chief, Big Joker, Black Hawk, Black Tail, Black Thorn, Captain, Cherokee Strip, Colorado, Dan Patch, Dead Shot, Etta, Good Ore, Greasy Run, Great Scott, Idora, IXL, Jay Bird, Kentucky Bell, King Solomon, Laurel, Laura S, Little Jay, Meteor, Neglected, Northern Light, Old Nell, Plymouth Rock, Polaris, Quandary, Reserve, Shoo Fly, Silver Plume, Three Pines, Vernie, White Rose and dozens of others. **8.5" X 11", 114 ppgs, Retail Price: $10.99**

The Index Mining District of Washington - Unavailable since 1912, this important publication was originally published by the Washington Geologic Survey and has been unavailable for a century. Topics include the geology, rock formations and the formation of ore deposits in this important mining area of Washington State. Also included are hard to find details on the geology, history and locations of dozens of mines in the area. Some of the mines featured include the Sunset, Non-Pareil, Ethel Consolidated, Kittaning, Merchant, Homestead, Co-operative, Lost Creek, Uncle Sam, Calumet, Florence-Rae, Bitter Creek, Index Peacock, Gunn Peak, Helena, North Star, Buckeye. Copper Bell, Red Cross and others. **8.5" X 11", 114 ppgs, Retail Price: $11.99**

Mining & Mineral Resources of Stevens County Washington - Unavailable since 1920, this important publication was originally published by the Washington Geologic Survey and has been unavailable for a century. Topics include the geology, rock formations and the formation of ore deposits in these important mining areas of Washington State. Also included are hard to find details on the geology, history and locations of hundreds of mines in the area. **8.5" X 11", 372 ppgs, Retail Price: $24.99**

The Mines and Geology of the Loomis Quadrangle Okanogan County, Washington - Unavailable since 1972, this important publication was originally published by the Washington Geologic Survey and has been unavailable for a century. Topics include the geology, rock formations and the formation of ore deposits in this important mining area of Washington State. Also included are hard to find details on the geology, history and locations of dozens of gold, copper, silver and other mines in the area. **8.5" X 11", 150 ppgs, Retail Price: $12.99**

The Conconully Mining District of Okanogan County Washington - Unavailable since 1973, this important publication was originally published by the Washington Geologic Survey and has been unavailable for a century. Topics include the geology, rock formations and the formation of ore deposits in this important mining area of Washington State, which also includes Salmon Creek, Blue Lake and Galena. Also included are hard to find details on the geology, mining history and locations of dozens of mines in the area. Some of the mines include Arlington, Fourth of July, Sonny Boy, First Thought, Last Chance, War Eagle-Peacock, Wheeler, Mohawk, Lone Star, Woo Loo Moo Loo, Keystone, Hughes, Plant-Callahan, Johnny Boy, Leuena, Gubser, John Arthur, Tough Nut, Homestake, Key and many others **8.5" X 11", 68 ppgs, Retail Price: $8.99**

Wyoming Mining Books

Mining in the Laramie Basin of Wyoming - Unavailable since 1909, this publication was originally compiled by the United States Department of Interior. Also included are insights into the mineralization and other characteristics of this important mining region, especially in regards to coal, limestone, gypsum, bentonite clay, cement, sand, clay and copper. **8.5" X 11", 104 ppgs, Retail Price: $11.99**

New Mexico Mining Books

The Mogollon Mining District of New Mexico - Unavailable since 1927, this important publication was originally published by the US Department of Interior and has been unavailable for 80 years. Topics include the geology, rock formations and the formation of ore deposits in this important mining area in New Mexico. Of particular focus is information on the history and production of the ore deposits in this area, their form and structure, vein filling, their paragenesis, origins and ore shoots, as well as oxidation and supergene enrichment. Also included are hard to find details, including the descriptions and locations of numerous gold, silver and other types of mines, including the Eureka, Pacific, South Alpine, Great Western, Enterprise, Buffalo, Mountain View, Floride, Gold Dust, Last Chance, Deadwood, Confidence, Maud S., Deep Down, Little Fanney, Trilby, Johnson, Alberta, Comet, Golden Eagle, Cooney, Queen, the Iron Crown, Eberle, Clifton, Andrew Jackson mine, Mascot and others. **8.5" X 11", 144 ppgs, Retail Price: $12.99**

The Percha Mining District of Kingston New Mexico - Unavailable since 1883, this important publication was originally published by the Kingston Tribune and has been unavailable for over one hundred and thirty five years. Having been written during the earliest years of gold and silver mining in the Percha Mining District, unlike other books on the subject, this work offers the unique perspective of having actually been written while the early mining history of this area was still being made. In fact, the work was written so early in the development of this area that many of the notable mines in the Percha District were less than a few years old and were still being operated by their original discoverers with the same enthusiasm as when they were first located. Included are hard to find details on the very earliest gold and silver mines of this important mining district near Kingston in Sierra County, New Mexico. **8.5" X 11", 68 ppgs, Retail Price: $9.99**

East Coast Mining Books

The Gold Fields of the Southern Appalachians - Unavailable since 1895, this important publication was originally published by the US Department of Interior and has been unavailable for nearly 120 years. Topics include the geology, rock formations and the formation of ore deposits in this important mining area of the American South. Of particular focus is information on the history and statistics of the ore deposits in this area, their form and structure and veins. Also included are details on the placer gold deposits of the region. The gold fields of the Georgian Belt, Carolinian Belt and the South Mountain Mining District of North Carolina are all treated in descriptive detail. Included are hard to find details, including the descriptions and locations of numerous gold mines in Georgia, North Carolina and elsewhere in the American South. Also included are details on the gold belts of the British Maritime Provinces and the Green Mountains. 8.5" X 11", 104 ppgs, **Retail Price: $9.99**

Gold Rush Tales Series

Millions in Siskiyou County Gold - In this first volume of the "Gold Rush Tales" series, leading mining historian and editor Kerby Jackson, introduces us to the story of how millions of dollars worth of gold was discovered in Siskiyou County during the California Gold Rush. Lavishly illustrated with photos from the 19th Century, this hard to find information was first published in 1897 and sheds important light onto the gold rush era in Siskiyou County, California and the experiences of the men who dug for the gold and actually found it. 8.5" X 11", 82 ppgs, **Retail Price: $9.99**

The California Rand in the Days of '49 - In this second volume of the "Gold Rush Tales" series, leading mining historian and editor Kerby Jackson, introduces us to four tales from the California Gold Rush. Lavishly illustrated with photos from the 19th Century, this hard to find information was first published in 1890's and includes the stories of "California's Rand", details about Chinese miners, how one early miner named Baker struck it rich and also the story of Alphonzo Bowers, who invented the first hydraulic gold dredge. 8.5" X 11", 54 ppgs, **Retail Price: $9.99**

More Mining Books

Prospecting and Developing A Small Mine - Topics covered include the classification of varying ores, how to take a proper ore sample, the proper reduction of ore samples, alluvial sampling, how to understand geology as it is applied to prospecting and mining, prospecting procedures, methods of ore treatment, the application of drilling and blasting in a small mine and other topics that the small scale miner will find of benefit. 8.5" X 11", 112 ppgs, **Retail Price: $11.99**

Timbering For Small Underground Mines - Topics covered include the selection of caps and posts, the treatment of mine timbers, how to install mine timbers, repairing damaged timbers, use of drift supports, headboards, squeeze sets, ore chute construction, mine cribbing, square set timbering methods, the use of steel and concrete sets and other topics that the small underground miner will find of benefit. This volume also includes twenty eight illustrations depicting the proper construction of mine timbering and support systems that greatly enhance the practical usability of the information contained in this small book. 8.5" X 11", 88 ppgs. **Retail Price: $10.99**

Timbering and Mining - A classic mining publication on Hard Rock Mining by W.H. Storms. Unavailable since 1909, this rare publication provides an in depth look at American methods of underground mine timbering and mining methods. Topics include the selection and preservation of mine timbers, drifting and drift sets, driving in running ground, structural steel in mine workings, timbering drifts in gravel mines, timbering methods for driving shafts, positioning drill holes in shafts, timbering stations at shafts, drainage, mining large ore bodies by means of open cuts or by the "Glory Hole" system, stoping out ore in flat or low lying veins, use of the "Caving System", stoping in swelling ground, how to stope out large ore bodies, Square Set timbering on the Comstock and its modifications by California miners, the construction of ore chutes, stoping ore bodies by use of the "Block System", how to work dangerous ground, information on the "Delprat System" of stoping without mine timbers, construction and use of headframes and much more. This volume provides a reference into not only practical methods of mining and timbering that may be employed in narrow vein mining by small miners today, but also rare insights into how mines were being worked at the turn of the 19th Century. 8.5" X 11", 288 ppgs. **Retail Price: $24.99**

A Study of Ore Deposits For The Practical Miner - Mining historian Kerby Jackson introduces us to a classic mining publication on ore deposits by J.P. Wallace. First published in 1908, it has been unavailable for over a century. Included are important insights into the properties of minerals and their identification, on the occurrence and origin of gold, on gold alloys, insights into gold bearing sulfides such as pyrites and arsenopyrites, on gold bearing vanadium, gold and silver tellurides, lead and mercury tellurides, on silver ores, platinum and iridium, mercury ores, copper ores, lead ores, zinc ores, iron ores, chromium ores, manganese ores, nickel ores, tin ores, tungsten ores and others. Also included are facts regarding rock forming minerals, their composition and occurrences, on igneous, sedimentary, metamorphic and intrusive rocks, as well as how they are geologically disturbed by dikes, flows and faults, as well as the effects of these geologic actions and why they are important to the miner. Written specifically with the common miner and prospector in mind, the book will help to unlock the earth's hidden wealth for you and is written in a simple and concise language that anyone can understand. 8.5″ X 11″, 366 ppgs. **Retail Price: $24.99**

Mine Drainage - Unavailable since 1896, this rare publication provides an in depth look at American methods of underground mine drainage and mining pump systems. This volume provides a reference into not only practical methods of mining drainage that may be employed in narrow vein mining by small miners today, but also rare insights into how mines were being worked at the turn of the 19th Century. 8.5″ X 11″, 218 ppgs. **Retail Price: $24.99**

Fire Assaying Gold, Silver and Lead Ores - Unavailable since 1907, this important publication was originally published by the Mining and Scientific Press and was designed to introduce miners and prospectors of gold, silver and lead to the art of fire assaying. Topics include the fire assaying of ores and products containing gold, silver and lead; the sampling and preparation of ore for an assay; care of the assay office, assay furnaces; crucibles and scorifiers; assay balances; metallic ores; scorification assays; cupelling; parting' crucible assays, the roasting of ores and more. This classic provides a time honored method of assaying put forward in a clear, concise and easy to understand language that will make it a benefit to even beginners. 8.5″ X 11″, 96 ppgs. **Retail Price: $11.99**

Methods of Mine Timbering - Originally published in 1896, this important publication on mining engineering has not been available for nearly a century. Included are rare insights into historical methods of timbering structural support that were used in underground metal mines during the California that still have a practical application for the small scale hardrock miner of today. 8.5″ X 11″, 94 ppgs. **Retail Price: $10.99**

The Enrichment of Copper Sulfide Ores - First published in 1913, it has been unavailable for over a century. Topics include the definition and types of ore enrichment, the oxidation of copper ores, the precipitation of metallic sulfides. Also included are the results of dozens of lab experiments pertaining to the enrichment of sulfide ores that will be of interest to the practical hard rock mine operator in his efforts to release the metallic bounty from his mine's ore. 8.5″ X 11″, 92 ppgs. **Retail Price: $9.99**

A Study of Magmatic Sulfide Ores - Unavailable since 1914, this rare publication provides an in depth look at magmatic sulfide ores. Some of the topics included are the definition and classification of magmatic ores, descriptions of some magmatic sulfide ore deposits known at the time of publication including copper and nickel bearing pyrrohitic ore bodies, chalcopyrite-bornite deposits, pyritic deposits, magnetite-ileminite deposits, chromite deposits and magmatic iron ore deposits. Also included are details on how to recognize these types of ore deposits while prospecting for valuable hardrock minerals. 8.5″ X 11″, 138 ppgs. **Retail Price: $11.99**

The Cyanide Process of Gold Recovery - Unavailable since 1894 and released under the name "The Cyanide Process: Its Practical Application and Economical Results", this rare publication provides an in depth look at the early use of cyanide leaching for gold recovery from hardrock mine ores. This volume provides a reference into the early development and use of cyanide leaching to recover gold. 8.5″ X 11″, 162 ppgs. **Retail Price: $14.99**

California Gold Milling Practices - Unavailable since 1895 and released under the name "California Gold Practices", this rare publication provides an in depth look at early methods of milling used to reduce gold ores in California during the late 19th century. This volume provides a reference into the early development and use of milling equipment during the earliest years of the California Gold Rush up to the age of the Industrial Revolution. Much of the information still applies today and will be of use to small scale miners engaging in hardrock mining. 8.5″ X 11″, 104 ppgs. **Retail Price: $10.99**

Leaching Gold and Silver Ores With The Plattner and Kiss Processes - Mining historian Kerby Jackson introduces us to a classic mining publication on the evaluation and examination of mines and prospects by C.H. Aaron. First published in 1881, it has been unavailable for over a century and sheds important light on the leaching of gold and silver ores with the Plattner and Kiss processes. 8.5″ X 11″, 204 ppgs. **Retail Price: $15.99**

The Metallurgy of Lead and the Desilverization of Base Bullion - First published in 1896, it has been unavailable for over a century and sheds important light on the the recovery of silver from lead based ores. Some of the topics include the properties of lead and some of its compounds, lead ores such as galenite, anglesite, cerussite and others, the distribution of lead ores throughout the United States and the sampling and assaying of lead ores. Also covered is the metallurgical treatment of lead ores, as well as the desilverization of lead by the Pattinson Process and the Parkes Process. Hofman's text has long been considered one of the most important early works on the recovery of silver from lead based ores. 8.5" X 11", 452 ppgs. Retail Price: $29.99

Ore Sampling For Small Scale Miners - First published in 1916, it has been unavailable for over a century and sheds important light on historic methods of ore sampling in hardrock mines. Topics include how to take correct ore samples and the conditions that affect sampling, such as their subdivision and uniformity. Particular detail is given to methods of hand sampling ore bodies by grab sample, pipe sample and coning, as well as sampling by mechanical methods. Also given are insights into the screening, drying and grinding processes to achieve the most consistent sample results and much more. 8.5" X 11", 124 ppgs. Retail Price: $12.99

The Extraction of Silver, Copper and Tin from Ores - First published in 1896, it has been unavailable for over a century and sheds important light on how historic miners recovered silver, copper and tin from their mining operations. The book is split into three sections, including a discussion on the Lixiviation of Silver Ores, the mining and treatment of copper ores as practiced at Tharsis, Spain and the smelting of tin as it was practiced by metallurgists at Pulo Brani, Singapore. Also included is an overview and analysis of these historic metal recovery methods that will be of benefit to those interested in the extraction of silver, copper and tin from small mines. 8.5" X 11", 118 ppgs. Retail Price: $14.99

The Roasting of Gold and Silver Ores - First published in 1880, it has been unavailable for over a century and sheds important light on how historic miners recovered gold and silver rom their mining operations. Topics include details on the most important silver and free milling gold ores, methods of desulphurization of ores, methods of deoxidation, the chlorination of ores, methods and details on roasting gold and silver ores, notes on furnaces and more. Also included are details on numerous methods of gold and silver recovery, including the Ottokar Hofman's Process, the Patera Process, Kiss Process, Augustin Process, Ziervogel Process and others. 8.5" X 11", 178 ppgs. Retail Price: $19.99

The Examination of Mines and Prospects - First published in 1912, it has been unavailable for over a century and sheds important light on how to examine and evaluate hardrock mines, prospects and lode mining claims. Sections include Mining Examinations, Structural Geology, Structural Features of Ore Deposits, Primary Ores and their Distribution, Types of Primary Ore Deposits, Primary Ore Shoots, The Primary Alteration of Wall Rocks, Alterations by Surface Agencies, Residual Ores and their Distribution, Secondary Ores and Ore Shoots and Vein Outcrops. This hard to find information is a must for those who are interested in owning a mine or who already own a lode mining claim and wish to succeed at quartz mining. 8.5" X 11", 250 ppgs. Retail Price: $19.99

Garnets: Their Mining, Milling and Utilization - First published in 1925, it has been unavailable since those days and sheds important light on the mining, milling and utilization of garnets. Included are details on the characteristics of garnets, where they are found and how they were mined. 78 ppgs, 10.99

Gemstones and Precious Stones of North America - Leading mining historian Kerby Jackson introduces us to a classic mining publication on the gems and precious stones of the United States, Canada and mexico. First published in 1890, it has been unavailable since those days and sheds important light on the gems and precious stones that may be found in North America. Included are chapters on diamonds, corundum, sapphire, ruby, topaz, emerald, disapore, spinel, turquoise, tourmaline, garnets, beyrl, peridot, zircon, quartz crystals, feldspars, pearls and many others. Included are details on where these gems and precious stones may be found throughout North America, as well as their characteristics. 360 ppgs, 24.99

Mining Camps and Mining Districts - First released in 1885 by Charles Howard Shinn under the title "Mining Camps: A Study in American Frontier Government", this publication offers a unique look at how early gold miners established their own forms of representative government during the California Gold Rush. Drawing on the the early mining codes of mideviel German miners in the Harz Mountains, on the mining customs of the Cornish tin miners and early Spanish mining laws introduced into California, the miners established the first governments in the American West. 340 ppgs, 24.99

BLM Field Handbook for Mineral Examiners - Leading mining historian Kerby Jackson introduces us to a classic mining publication on mine evaluation. First published in 1962, this work sheds important light on the techniques of BLM Mineral Examiners to perform validity on mining claims. 132 ppgs, 10.99

<u>**Six Months In The Gold Mines During The California Gold Rush**</u> - Unavailable since 1850, this important work is a first hand account of one "49'ers" personal experience during the great California Gold Rush, shedding important light on one of the most exciting periods in the history of not only California, but also the world. Compiled from journals written between 1847 and 1849 by E. Gould Buffum, a native of New York, "Six Months In The Gold Mines During The California Gold Rush" offers a rare look into the day to day lives of the people who came to California to work in her gold mines when the state was still a great frontier. **8.5" X 11", 290 ppgs. Retail Price: $19.99**

<u>**The Discovery of Gold in Australia**</u> - First published in 1852, it has been unavailable since those days and sheds important light on Australia's gold mining history. Included are rare communications between British agents and the British Crown when gold was first discovered in Australia in 1851. This rare text contains hard to find details on Australia's first mining camps and Britain's early attempts to provide for the orderly regulation of gold mines in that part of the world. Also of interest are hard to find extracts of articles that appeared in the early colonial newspapers that did their best to report on Australia's gold rush as it took place.
102 ppgs, 10.99

www.ingramcontent.com/pod-product-compliance
Lightning Source LLC
Chambersburg PA
CBHW080608180526
45168CB00007B/2823